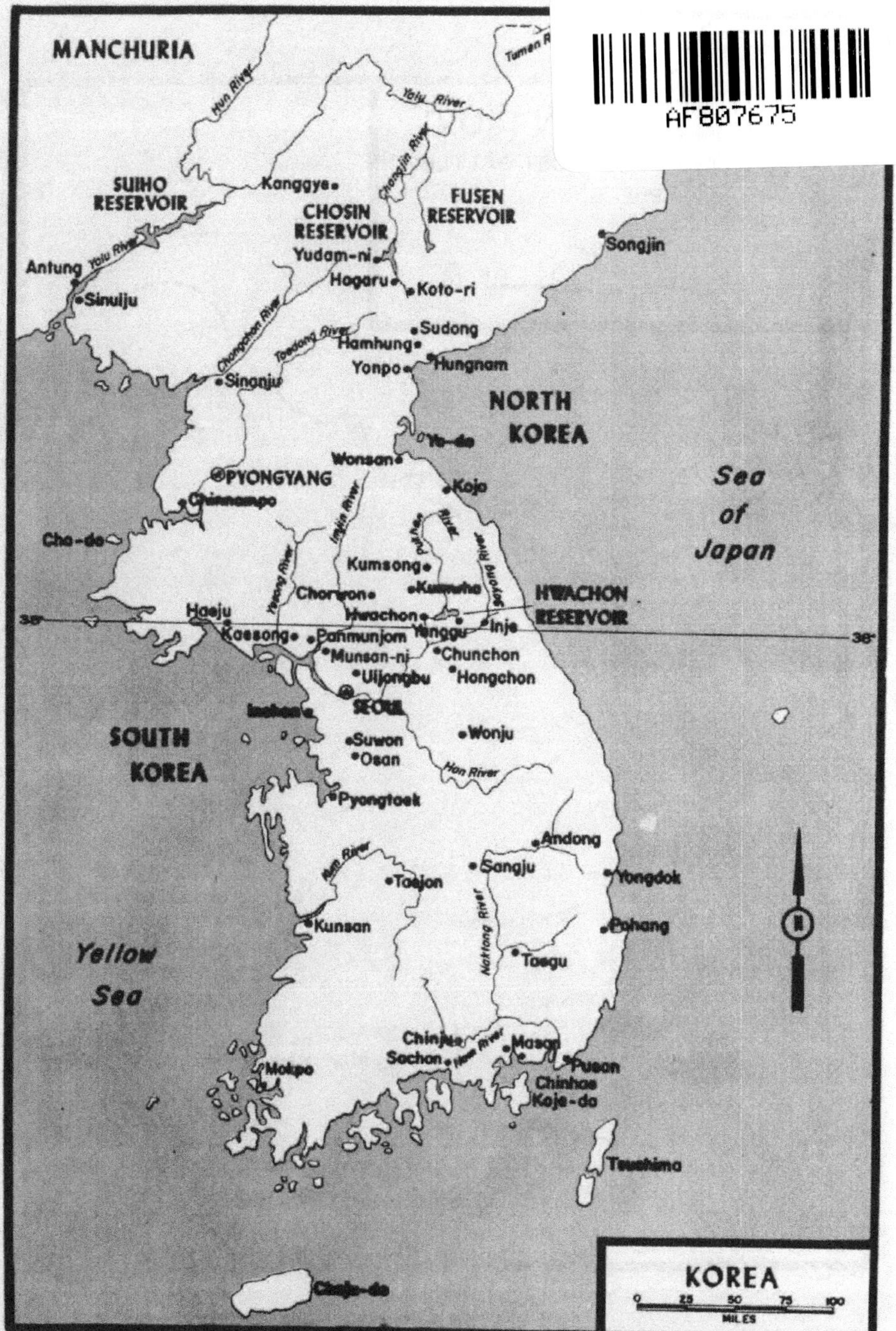
MANCHURIA
Hun River
Yalu River
Tumen R
Changjin River
SUIHO
RESERVOIR
Kanggye
CHOSIN
RESERVOIR
FUSEN
RESERVOIR
Yudam-ni
Antung
Yalu River
Hagaru
Koto-ri
Sinuiju
Songjin
Chongchon River
Taedong River
Sudong
Hamhung
Yonpo
Hungnam
Sinanju
NORTH
KOREA
Wonsan
PYONGYANG
Kojo
Imjin River
Sea
of
Japan
Cho-do
Kumsong
Yesong River
Soyang River
Chorwon
HWACHON
RESERVOIR
Haeju
Hwachon
Inje
38°
Kaesong
Panmunjom
Yanggu
Munsan-ni
Chunchon
Uijongbu
Hongchon
SEOUL
SOUTH
KOREA
Suwon
Wonju
Osan
Han River
Pyongtaek
Andong
Kum River
Sangju
Taejon
Yongdok
Kunsan
Naktong River
Pohang
Yellow
Sea
Taegu
Chinju
Masan
Sachon
Nam River
Pusan
Mokpo
Chinhae
Koje-do
Tsushima
KOREA
0
25
50
75
100
MILES

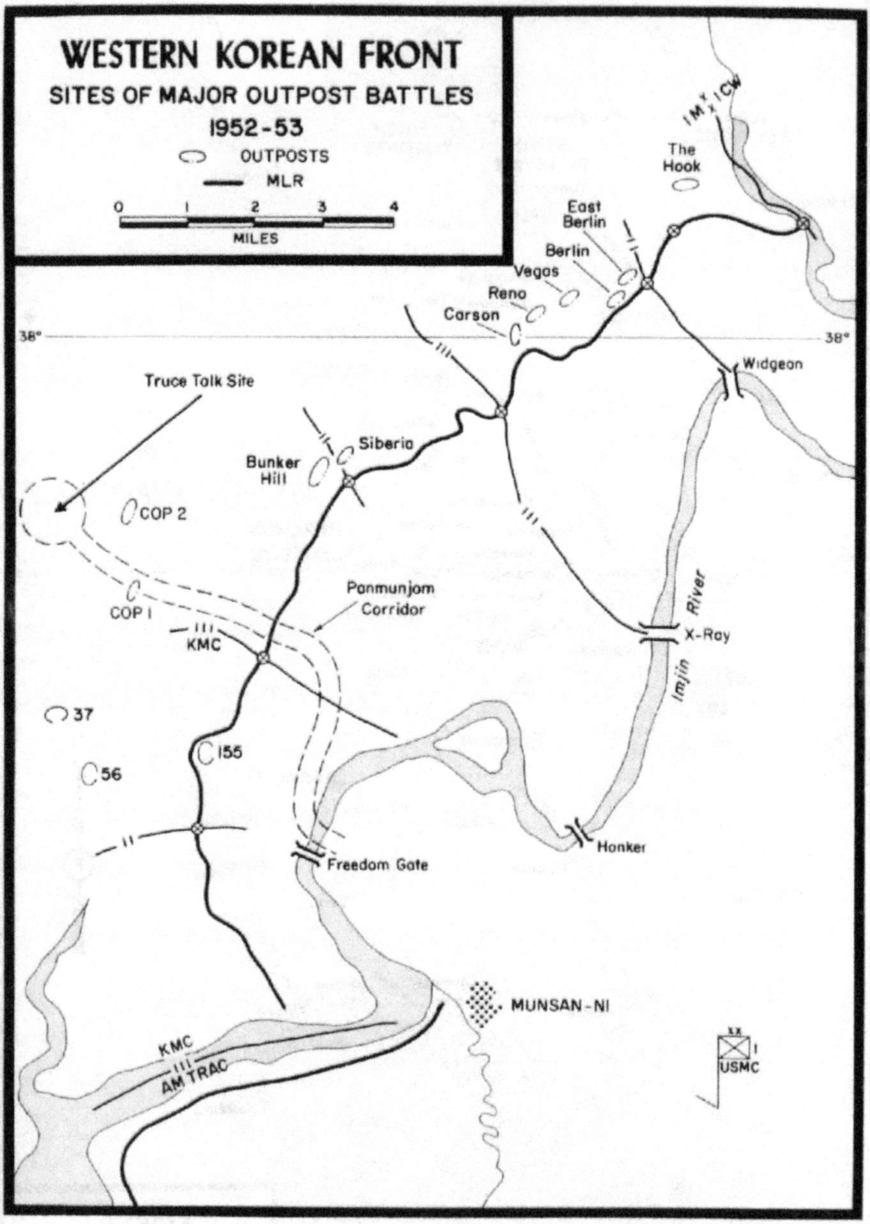

WESTERN KOREAN FRONT
SITES OF MAJOR OUTPOST BATTLES
1952-53
OUTPOSTS
MLR
0 1 2 3 4
MILES
The Hook
East Berlin
Berlin
Vegas
Reno
Carson
38°
38°
Widgeon
Truce Talk Site
Siberia
Bunker Hill
COP 2
COP 1
Panmunjom Corridor
KMC
X-Ray
Imjin River
37
155
56
Honker
Freedom Gate
MUNSAN-NI
KMC
AM TRAC
USMC

U. S. MARINE OPERATIONS IN KOREA

1950–1953

VOLUME V

Operations in West Korea

by

LIEUTENANT COLONEL PAT MEID, USMCR

and

MAJOR JAMES M. YINGLING, USMC

Historical Division

Headquarters, U. S. Marine Corps

Washington, D. C., 1972

Preceding Volumes of

U. S. Marine Operations in Korea

Volume I, "The Pusan Perimeter"

Volume II, "The Inchon-Seoul Campaign"

Volume III, "The Chosin Reservoir Campaign"

Volume IV, "The East-Central Front"

Foreword

MENTION THE KOREAN WAR and almost immediately it evokes the memory of Marines at Pusan, Inchon, Chosin Reservoir, or the Punchbowl. Americans everywhere remember the Marine Corps' combat readiness, courage, and military skills that were largely responsible for the success of these early operations in 1950–1951. Not as dramatic or well-known are the important accomplishments of the Marines during the latter part of the Korean War.

In March 1952 the 1st Marine Division redeployed from the East-Central front to West Korea. This new sector, nearly 35 miles in length, anchored the far western end of I Corps and was one of the most critical of the entire Eighth Army line. Here the Marines blocked the enemy's goal of penetrating to Seoul, the South Korean capital. Northwest of the Marine Main Line of Resistance, less than five miles distant, lay Panmunjom, site of the sporadic truce negotiations.

Defense of their strategic area exposed the Marines to continuous and deadly Communist probes and limited objective attacks. These bitter and costly contests for key outposts bore such names as Bunker Hill, the Hook, the Nevadas (Carson-Reno-Vegas), and Boulder City. For the ground Marines, supported by 1st Marine Aircraft Wing squadrons, the fighting continued until the last day of the war, 27 July 1953.

The Korean War marked the first real test of Free World solidarity in the face of Communist force. In repulsing this attempted Communist aggression, the United Nations, led by the United States, served notice that it would not hesitate to aid those nations whose freedom and independence were under attack.

As events have subsequently proven, holding the line against Communist encroachment is a battle whose end is not yet in sight. Enemy aggression may explode brazenly upon the world scene, with an overt act of invasion, as it did in Korea in June 1950, or it may take the form of a murderous guerrilla war as it has more recently, for over a decade, in Vietnam.

Whatever guise the enemy of the United States chooses or wherever he draws his battleline, he will find the Marines with their age-old answer. Today, as in the Korean era, Marine Corps readiness and professionalism are prepared to apply the cutting edge against any threat to American security.

L. F. CHAPMAN, JR.
General, U.S. Marine Corps,
Commandant of the Marine Corps

Reviewed and approved: 12 May 1971.

Preface

THIS IS THE CONCLUDING VOLUME of a five-part series dealing with operations of United States Marines in Korea between 2 August 1950 and 27 July 1953. Volume V provides a definitive account of operations of the 1st Marine Division and the 1st Marine Aircraft Wing during 1952–1953, the final phase of the Korean War. At this time the division operated under Eighth U.S. Army in Korea (EUSAK) control in the far western sector of I Corps, while Marine aviators and squadrons functioned as a component of the Fifth Air Force (FAF).

The period covered by this history begins in March 1952, when the Marine division moved west to occupy positions defending the approaches to Seoul, the South Korean capital. As it had for most of the war the 1st Marine Aircraft Wing, operating under FAF, flew close support missions not only for the Marines but for as many as 19 other Allied frontline divisions. Included in the narrative is a detailed account of Marine POWs, a discussion of the new defense mission of Marine units in the immediate postwar period, and an evaluation of Marine Corps contributions to the Korean War.

Marines, both ground and aviation, comprised an integral part of the United Nations Command in Korea. Since this is primarily a Marine Corps history, actions of the U.S. Army, Navy, and Air Force are presented only in sufficient detail to place Marine operations in their proper perspective.

Official Marine Corps combat records form the basis for the book. This primary source material has been further supplemented by comments and interviews from key participants in the action described. More than 180 persons reviewed the draft chapters. Their technical knowledge and advice have been invaluable. Although the full details of these comments could not be used in the text, this material has been placed in Marine Corps archives for possible use by future historians.

The manuscript of this volume was prepared during the tenure of Colonel Frank C. Caldwell, Director of Marine Corps History,

Historical Division, Headquarters Marine Corps. Production was accomplished under the direction of Mr. Henry I. Shaw, Jr., Deputy Director and Chief Historian, who also outlined the volume. Preliminary drafts were written by the late Lynn Montross, prime author of this series, and Major Hubard D. Kuokka. Major James M. Yingling researched and wrote chapters 1–6 and compiled the Command and Staff List. Lieutenant Colonel Pat Meid researched and wrote chapters 7–12, prepared appendices, processed photographs and maps, and did the final editing of the book.

Historical Division staff members, past or present, who freely lent suggestions or provided information include Lieutenant Colonel John J. Cahill, Captain Charles B. Collins, Mr. Ralph W. Donnelly, Mr. Benis M. Frank, Mr. George W. Garand, Mr. Rowland P. Gill, Captain Robert J. Kane, Major Jack K. Ringler, and Major Lloyd E. Tatem. Warrant Officer Dennis Egan was Administrative Officer during the final stages of preparation and production of this book.

The many exacting administrative duties involved in processing the volume from first draft manuscripts through the final printed form, including the formidable task of indexing the book, were handled expertly and cheerfully by Miss Kay P. Sue. Mrs. Frances J. Rubright also furnished gracious and speedy assistance in obtaining the tomes of official Marine Corps records. The maps were prepared by Sergeants Kenneth W. White and Ernest L. Wilson. Official Department of Defense photographs illustrate the book.

A major contribution to the history was made by the Office of the Chief of Military History, Department of the Army; the Naval History Division, Department of the Navy; and the Office of Air Force History, Department of the Air Force. Military history offices of England, Canada, and South Korea provided additional details that add to the accuracy and interest of this concluding volume of the Korean series.

F. C. Caldwell
Colonel, U.S. Marine Corps (Retired)
Director of Marine Corps History

Contents

Appendices

Illustrations

Photographs

Sixteen-page sections of photographs following pages 212 and 436

Maps and Sketches

Illustrations

CHAPTER I

Operations in West Korea Begin

From Cairo to JAMESTOWN—*The Marines' Home in West Korea—Organization of the 1st Marine Division Area—The 1st Marine Aircraft Wing—The Enemy—Initial CCF Attack —Subsequent CCF Attacks—Strengthening the Line—Marine Air Operations—Supporting the Division and the Wing— Different Area, Different Problem*

From Cairo to JAMESTOWN[1]

DURING THE LATTER PART of March 1952, the 1st Marine Division, a component of the U.S. Eighth Army in Korea (EUSAK), pulled out of its positions astride the Soyang River in east-central Korea and moved to the far western part of the country in the I Corps sector. There the Marines took over the EUSAK left flank, guarding the most likely enemy approaches to the South Korean capital city, Seoul, and improving the ground defense in their sector to comply with the strict requirements which the division

[1] Unless otherwise noted, the material in this section is derived from: 1st Marine Division Staff Report, titled "Notes for Major General J. T. Selden, Commanding General, First Marine Division, Korea," dtd 20 Aug 52, hereafter Selden, *Div. Staff Rpt;* the four previous volumes of the series *U.S. Marine Operations in Korea, 1950–1953,* namely, Lynn Montross and Capt Nicholas A. Canzona, *The Pusan Perimeter,* v. I; *The Inchon-Seoul Operation,* v. II; *The Chosin Reservoir Campaign,* v. III; Lynn Montross, Maj Hubard D. Kuokka, and Maj Norman W. Hicks, *The East-Central Front,* v. IV (Washington: HistBr, G–3 Div, HQMC, 1954–1962), hereafter Montross, Kuokka, and Hicks, *USMC Ops Korea—Central Front,* v. IV; Department of Military Art and Engineering, U.S. Military Academy, *Operations in Korea* (West Point, N.Y.: 1956), hereafter USMA, *Korea;* David Rees, *Korea: The Limited War* (New York: St. Martin's Press, 1964), hereafter Rees, *Korea,* quoted with permission of the publisher. Unless otherwise noted, all documentary material cited is on file at, or obtainable through, the Archives of the Historical Division, Headquarters, U.S. Marine Corps.

commander, Major General John T. Selden, had set down. Except for a brief period in reserve, the Marine division would remain in the Korean front lines until a cease-fire agreement in July 1953 ended active hostilities.

The division CG, Major General Selden,[2] had assumed command of the 25,000-man 1st Marine Division two months earlier, on 11 January, from Major General Gerald C. Thomas while the Marines were still in the eastern X Corps sector. The new Marine commander was a 37-year veteran of Marine Corps service, having enlisted as a private in 1915, serving shortly thereafter in Haiti. During World War I he was commissioned a second lieutenant, in 1918, while on convoy duty. Between the two world wars, his overseas service had included a second assignment to Haiti, two China tours, and sea duty. When the United States entered World War II, Lieutenant Colonel Selden was an intelligence officer aboard the carrier *Lexington.* Later in the war Colonel Selden led the 5th Marines in the New Britain fighting and was Chief of Staff of the 1st Marine Division in the Peleliu campaign. He was promoted to brigadier general in 1948 and received his second star in 1951, prior to his combat assignment in Korea.

American concern in the 1950s for South Korea's struggle to preserve its independence stemmed from a World War II agreement between the United States, the United Kingdom, and China. In December 1943, the three powers had signed the Cairo Declaration and bound themselves to ensure the freedom of the Korean people, then under the yoke of the Japanese Empire. At the Potsdam Conference, held on the outskirts of Berlin, Germany in July 1945, the United States, China,[3] and Britain renewed their Cairo promise.

When the Soviet Union agreed to join forces against Japan, on 8 August, the USSR also became a party to the Cairo Declaration. According to terms of the Japanese capitulation on 11 August, the Soviets were to accept surrender of the defeated forces north of the 38th Parallel in Korea. South of that line, the commander of the

---

[2] DivInfo, HQMC, Biography of MajGen John T. Selden, Mar 54.

[3] China did not attend. Instead, it received an advance copy of the proposed text. President Chiang Kai-shek signified Chinese approval on 26 July. A few hours later, the Potsdam Declaration was made public. *Foreign Relations of the United States: The Conferences at Cairo and Teheran, 1943* (Department of State publication 7187), pp. 448–449; *The Conference of Berlin (The Potsdam Conference), 1945*, v. II (Department of State publication 7163), pp. 1278, 1282–1283, 1474–1476.

American occupation forces would receive the surrender. The Russians wasted no time and on 12 August had their troops in northern Korea. American combat units, deployed throughout the Pacific, did not enter Korea until 8 September. Then they found the Soviet soldiers so firmly established they even refused to permit U.S. occupation officials from the south to cross over into the Russian sector. A December conference in Moscow led to a Russo-American commission to work out the postwar problems of Korean independence.

Meeting for the first time in March 1946, the commission was short-lived. Its failure, due to lack of Russian co-operation, paved the way for politico-military factions within the country that set up two separate Koreas. In the north the Communists, under Kim Il Sung, and in the south the Korean nationalists, led by Dr. Syngman Rhee, organized independent governments early in 1947. In May of that year, a second joint commission failed to unify the country. As a result the Korean problem was presented to the United Nations (UN). This postwar international agency was no more successful in resolving the differences between the disputing factions. It did, however, recognize the Rhee government in December 1948 as the representative one of the two dissident groups.

In June 1950, the North Koreans attempted to force unification of Korea under Communist control by crossing the 38th Parallel with seven infantry divisions heavily supported by artillery and tanks. Acting on a resolution presented by the United States, the United Nations responded by declaring the North Korean action a "breach of the peace" and called upon its members to assist the South Koreans in ousting the invaders. Many free countries around the globe offered their aid. In the United States, President Harry S. Truman authorized the use of U.S. air and naval units and, shortly thereafter, ground forces to evict the aggressors and restore the status quo. Under the command of General of the Army Douglas MacArthur, then Far East Commander, U.S. Eighth Army occupation troops in Japan embarked to South Korea.

The first combat unit sent from America to Korea was a Marine air-ground team, the 1st Provisional Marine Brigade, formed at Camp Pendleton, California on 7 July 1950, under Brigadier General Edward A. Craig. The same day the UN Security Council passed a resolution creating the United Nations Command (UNC) which was to exercise operational control over the international military forces rallying to the defense of South Korea. The Council asked the United States to appoint a commander of the UN forces; on the 8th, President Truman named his Far East Commander, General MacArthur, as Commander in Chief, United Nations Command (CinCUNC).

In Korea the Marines soon became known as the firemen of the Pusan Perimeter, for they were shifted from one trouble spot to the next all along the defensive ring around Pusan, the last United Nations stronghold in southeastern Korea during the early days of the fighting. A bold tactical stroke planned for mid-September was designed to relieve enemy pressure on Pusan and weaken the strength of the North Korean People's Army (NKPA). As envisioned by General MacArthur, an amphibious landing at Inchon on the west coast, far to the enemy rear, would threaten the entire North Korean position south of the 38th Parallel. To help effect this coup, the UN Commander directed that the Marine brigade be pulled out of the Pusan area to take part in the landing at Inchon.

MacArthur's assault force consisted of the 1st Marine Division, less one of its three regiments,[4] but including the 1st Korean Marine Corps (KMC) Regiment. Marine ground and aviation units were to assist in retaking Seoul, the South Korean capital, and to cut the supply line sustaining the NKPA divisions.

On 15 September, Marines stormed ashore on three Inchon beaches. Despite difficulties inherent in effecting a landing there,[5] it was an outstandingly successful amphibious assault. The 1st and 5th Marines, with 1st Marine Aircraft Wing (1st MAW) assault squadrons providing close air support, quickly captured the port city of Inchon, Ascom City[6] to the east, and Kimpo Airfield. Advancing eastward the Marines approached the Han River that separates Kimpo Peninsula from the Korean mainland. Crossing this obstacle in amphibian vehicles, 1st Division Marines converged on Seoul from three directions. By 27 September, the Marines had captured

---

[4] The 7th Marines was on its way to Korea at the time of the Inchon landing. The brigade, however, joined the 1st Division at sea en route to the objective to provide elements of the 5th Regimental Combat Team (RCT).

[5] For a discussion of the hardships facing the landing force, see Montross and Canzona, *USMC Ops Korea—Inchon,* v. II, *op. cit.,* pp. 41–42, 59–60, 62–64.

[6] In World War II, the Japanese developed a logistical base east of Inchon. When the Japanese surrendered, the Army Service Command temporarily took over the installation, naming it Ascom City. Maj Robert K. Sawyer, *Military Advisers in Korea: KMAG in Peace and War* (Washington: OCMH, DA, 1962), p. 43*n*.

the South Korean government complex and, together with the U.S. Army 7th Infantry Division, had severed the enemy's main supply route (MSR) to Pusan. In heavy, close fighting near the city, other United Nations troops pursued and cut off major units of the NKPA.

Ordered back to East Korea, the Marine division re-embarked at Inchon in October and made an administrative landing at Wonsan on the North Korean coast 75 miles above the 38th Parallel. As part of the U.S. X Corps, the 1st Marine Division was to move the 5th and 7th Marines (Reinforced) to the vicinity of the Chosin Reservoir, from where they were to continue the advance northward toward the North Korean-Manchurian border. The 1st Marines and support troops were to remain in the Wonsan area.

While the bulk of the division moved northward, an unforeseen development was in the making that was to change materially the military situation in Korea overnight. Aware that the North Koreans were on the brink of military disaster, Communist China had decided to enter the fighting. Nine Chinese divisions had been dispatched into the area with the specific mission of destroying the 1st Marine Division.[7] Without prior warning, on the night of 27 November, hordes of Chinese Communist Forces (CCF, or "Chinese People's Volunteers" as they called themselves) assaulted the unsuspecting Marines and nearly succeeding in trapping the two Marine regiments. The enemy's failure to do so was due to the military discipline and courage displayed by able-bodied and wounded Marines alike, as well as effective support furnished by Marine aviation. Under conditions of great hardship, the division fought its way out over 78 miles of frozen ground from Chosin to the port of Hungnam, where transports stood by to evacuate the weary men and the equipment they had salvaged.

This Chinese offensive had wrested victory from the grasp of General MacArthur just as the successful completion of the campaign seemed assured. In the west, the bulk of the Eighth Army paced its withdrawal with that of the X Corps. The UNC established a major line of defense across the country generally following the 38th Parallel. On Christmas Day, massed Chinese forces crossed the parallel, and within a week the UN positions were bearing the full brunt of the enemy assault. Driving southward, the Communists

---

[7] Montross and Canzona, *USMC Ops Korea—Chosin*, v. III, p. 161.

recaptured Seoul, but by mid-February 1951 the advance had been slowed down, the result of determined Eighth Army stands from a series of successive defensive lines.[8]

Following its evacuation from Hungnam, the 1st Marine Division early in 1951 underwent a brief period of rehabilitation and training in the vicinity of Masan, west of Pusan. From there, the division moved northeast to an area beyond Pohang on the east coast. Under operational control of Eighth Army, the Marines, with the 1st Korean Marine Corps Regiment attached for most of the period, protected 75 miles of a vital supply route from attack by bands of guerrillas. In addition, the Marines conducted patrols to locate, trap, and destroy the enemy. The Pohang guerrilla hunt also provided valuable training for several thousand recently arrived Marine division replacements.

In mid-February the 1st Marine Division was assigned to the U.S. IX Corps, then operating in east-central Korea near Wonju. Initially without the KMCs,[9] the Marine division helped push the corps line across the 38th Parallel into North Korea. On 22 April, the Chinese unleashed a gigantic offensive, which again forced UN troops back into South Korea. By the end of the month, however, the Allies had halted the 40-mile-wide enemy spring offensive.

Once again, in May, the Marine division was assigned to the U.S. X Corps, east of the IX Corps sector. Shortly thereafter the Communists launched another major offensive. Heavy casualties inflicted by UNC forces slowed this new enemy drive. Marine, Army, and Korean troops not only repelled the Chinese onslaught but immediately launched a counteroffensive, routing the enemy back into North Korea until the rough, mountainous terrain and stiffening resistance conspired to slow the Allied advance.

In addition to these combat difficulties, the Marine division began to encounter increasing trouble in obtaining what it considered

---

[8] On 9 January 1951, General MacArthur was "directed to defend himself in successive positions, inflicting maximum damage to hostile forces in Korea subject to the primary consideration of the safety of his troops and his basic mission of protecting Japan." Carl Berger, *The Korea Knot—A Military-Political History* (Philadelphia: University of Pennsylvania Press, 1957), pp. 131–132, hereafter Berger, *Korea Knot,* quoted with permission of the publisher.

[9] The 1st KMC Regiment was again attached to the Marine Division on 17 March 1951 and remained under its operational control for the remainder of the war. CinCPacFlt Interim Evaluation Rpt No. 4, Chap 9, p. 9–53, hereafter *PacFlt EvalRpt* with number and chapter.

sufficient and timely close air support (CAS). Most attack and fighter aircraft of the 1st MAW, commanded by Major General Field Harris[10] and operating since the Chosin Reservoir days under Fifth Air Force (FAF), had been employed primarily in a program of interdicting North Korean supply routes. Due to this diversion of Marine air from its primary CAS mission, both the division and wing suffered—the latter by its pilots' limited experience in performing precision CAS sorties. Despite the difficulties, the Marine division drove northward reaching, by 20 June, a grotesque scooped-out terrain feature on the east-central front appropriately dubbed the Punchbowl.

Eighth Army advances into North Korea had caused the enemy to reappraise his military situation. On 23 June, the Russian delegate to the United Nations, Jacob Malik, hinted that the Korean differences might be settled at the conference table. Subsequently, United Nations Command and Communist leaders agreed that truce negotiations would begin on 7 July at Kaesong, located in West Korea immediately south of the 38th Parallel, but under Communist control. The Communists broke off the talks on 22 August. Without offering any credible evidence, they declared that UNC aircraft had violated the neutrality zone surrounding the conference area.[11] Military and political observers then realized that the enemy's overture to peace negotiations had served its intended purpose of permitting him to slow his retreat, regroup his forces, and prepare his ground defenses for a new determined stand.

The lull in military offensive activity during the mid-1951 truce talks presaged the kind of warfare that would soon typify the final phase of the Korean conflict. Before the fighting settled into positional trench warfare reminiscent of World War I, the Marines participated in the final UN offensive. In a bitter struggle, the division hacked its way northward through, over, and around the Punchbowl, and in September 1951 occupied a series of commanding ter-

---

[10] Command responsibility of 1st MAW changed on 29 May 51 when Brigadier General Thomas J. Cushman succeeded General Harris.

[11] The Senior Delegate and Chief of the United Nations Command Delegation to the Korean Armistice Commission, Vice Admiral C. Turner Joy, USN, has described how the Communists in Korea concocted incidents "calculated to provide advantage for their negotiating efforts or for their basic propaganda objectives, or for both." Examples of such duplicity are given in Chapter IV of his book, *How Communists Negotiate* (New York: The MacMillan Company, 1955), hereafter Joy, *Truce Negotiations*, quoted with permission of the publisher. The quote above appears on p. 30.

rain positions that became part of the MINNESOTA Line, the Eighth Army main defensive line. Beginning on the 20th of that month, it became the primary mission of frontline units to organize, construct, and defend positions they held on MINNESOTA. To show good faith at the peace table, the UNC outlawed large-scale attacks against the enemy. Intent upon not appearing the aggressor and determined to keep the door open for future truce negotiations, the United Nations Command in late 1951 decreed a new military policy of limited offensives and an aggressive defense of its line. This change in Allied strategy, due to politico-military considerations, from a moving battle situation to stabilized warfare would affect both the tactics and future of the Korean War.

Even as Allied major tactical offensive operations and the era of fire and maneuver in Korea was passing into oblivion, several innovations were coming into use. One was the Marine Corps employment of helicopters. First used for evacuation of casualties from Pusan in August 1950, the versatile aircraft had also been adopted by the Marine brigade commander, General Craig, as an airborne jeep. On 13 September 1951, Marines made a significant contribution to the military profession when they introduced helicopters for large-scale resupply combat operations. This mission was followed one week later by the first use of helicopters for a combat zone troop lift. These revolutionary air tactics were contemporary with two new Marine Corps developments in ground equipment—body armor and insulated combat boots, which underwent extensive combat testing that summer and fall. The latter were to be especially welcomed for field use during the 1951-1952 winter.

Along the MINNESOTA Line, neither the freezing cold of a Korean winter nor blazing summer heat altered the daily routine. Ground defense operations consisted of dispatching patrols and raiding parties, laying ambushes, and improving the physical defenses. The enemy seemed reluctant to engage UN forces, and on one occasion to draw him into the open, EUSAK ordered Operation CLAM-UP across the entire UN front, beginning 10 February. Under cover of darkness, reserve battalions moved forward; then, during daylight, they pulled back, simulating a withdrawal of the main defenses. At the same time, frontline troops had explicit orders not to fire or even show themselves.[12]

---

[12] Col Franklin B. Nihart comments on draft MS, Sep 66, hereafter *Nihart comments.*

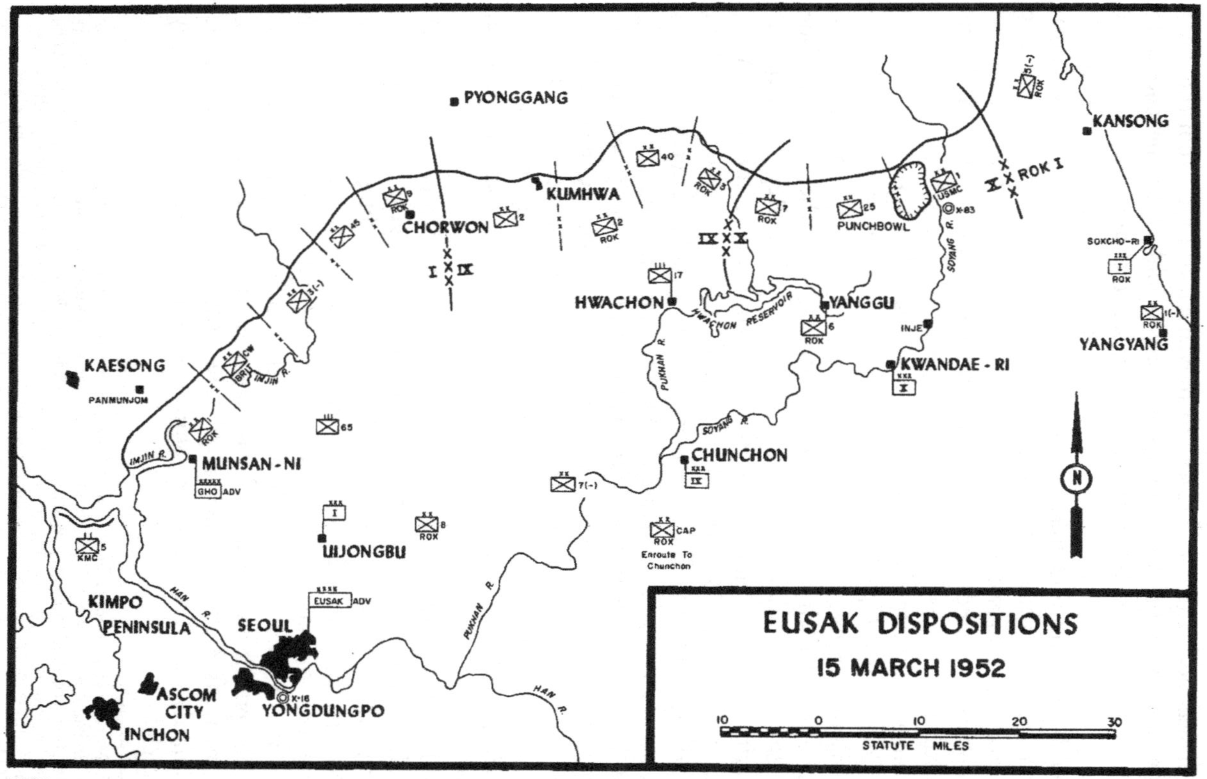

EUSAK DISPOSITIONS
15 MARCH 1952
STATUTE MILES
PYONGGANG
KANSONG
KUMHWA
CHORWON
PUNCHBOWL
ROK I
SOKCHO-RI
YANGYANG
HWACHON
HWACHON RESERVOIR
YANGGU
INJE
KWANDAE - RI
SOYANG R.
PUKHAN R.
CHUNCHON
KAESONG
PANMUNJOM
IMJIN R.
MUNSAN-NI
UIJONGBU
Enroute To Chunchon
KIMPO PENINSULA
SEOUL
HAN R.
ASCOM CITY
INCHON
YONGDUNGPO
EUSAK
GHQ

MAP 1

K WHITE

It was hoped that the rearward movement of units from the front line and the subsequent inactivity there would cause the enemy to come out of his trenches to investigate the apparent large-scale withdrawal of UNC troops. Then Marine and other EUSAK troops could open fire and inflict maximum casualties from covered positions. On the fifth day of the operation, CLAM-UP was ended. The North Koreans were lured out of their defenses, but not in the numbers expected. CLAM-UP was the last action in the X Corps sector for the 1st Marine Division, which would begin its cross-country relocation the following month. (See Map 1.)

Code-named Operation MIXMASTER, the transfer of the 1st Marine Division began on 17 March when major infantry units began to move out of their eastern X Corps positions, after their relief on line by the 8th Republic of Korea (ROK) Division. Regiments of the Marine division relocated in the following order: the 1st KMCs, 1st, 7th, and 5th Marines. The division's artillery regiment, the 11th Marines, made the shift by battalions at two-day intervals. In the motor march to West Korea, Marine units traveled approximately 140 miles over narrow, mountainous, and frequently mud-clogged primitive roads. Day and night, division transport augmented by a motor transport battalion attached from Fleet Marine Force, Pacific (FMFPac) and one company from the 1st Combat Service Group (CSG) rolled through rain, snow, sleet, and occasional good weather.

Marines employed 5,800 truck and DUKW (amphibious truck) loads to move most of the division personnel, gear, and supplies. Sixty-three flatbed trailers, 83 railroad cars, 14 landing ships, 2 transport aircraft, the vehicles of 4 Army truck companies, as well as hundreds of smaller jeep trailers and jeeps were utilized. The division estimated that these carriers moved about 50,000 tons of equipment and vehicles,[13] with some of the support units making as many as a dozen round trips. The MIXMASTER move was made primarily by truck and by ship[14] or rail for units with heavy vehicles.

---

[13] Marine commanders and staff officers involved in the planning and execution of the division move were alarmed at the amount of additional equipment that infantry units had acquired during the static battle situation. Many had become overburdened with "nice-to-have" items in excess of actual T/E (Table of Equipment) allowances. Col William P. Pala comments on draft MS, 5 Sep 66, hereafter *Pala comments*.

[14] Heavy equipment and tracked vehicles were loaded aboard LSDs and LSTs which sailed from Sokcho-ri to Inchon.

Impressive as these figures are, they almost pall in significance compared with the meticulous planning and precision logistics required by the week-long move. It was made, without mishap, over main routes that supplied nearly a dozen other divisions on the EUSAK line and thus had to be executed so as not to interfere with combat support. Although the transfer of the 1st Marine Division from the eastern to western front was the longest transplacement of any EUSAK division, MIXMASTER was a complicated tactical maneuver that involved realignment of UNC divisions across the entire Korean front. Some 200,000 men and their combat equipment had to be relocated as part of a master plan to strengthen the Allied front and deploy more troops on line.

Upon its arrival in West Korea, the 1st Marine Division was under orders to relieve the 1st ROK Division and take over a sector at the extreme left of the Eighth Army line, under I Corps control, where the weaknesses of Kimpo Peninsula defenses had been of considerable concern to EUSAK and its commander, General James A. Van Fleet. As division units reached their new sector, they moved to locations pre-selected in accordance with their assigned mission. First Marine unit into the I Corps main defensive position, the JAMESTOWN Line, was the 1st KMC Regiment attached to the division, with its organic artillery battalion. The KMCs, as well as 1/11, began to move into their new positions on 18 March. At 1400 on 20 March, the Korean Marines completed the relief of the 15th Republic of Korea Regiment in the left sector of the MLR (main line of resistance). Next into the division line, occupying the right regimental sector adjacent to the 1st Commonwealth Division, was Colonel Sidney S. Wade's 1st Marines with three battalions forward and 2/5 attached as the regimental reserve. Relief of the 1st ROK Division was completed on the night of 24-25 March. At 0400 on 25 March the Commanding General, 1st Marine Division assumed responsibility for the defense of 32 miles of the JAMESTOWN Line. That same date the remainder of the Marine artillery battalions also relocated in their new positions.

As the division took over its new I Corps mission on 25 March, the Marine commander had one regiment of the 1st ROK Division attached as division reserve while his 5th Marines was still in the east. Operational plans originally had called for the 5th Marines, less a battalion, to locate in the Kimpo Peninsula area where it was anticipated Marine reserve units would be able to conduct extensive amphibious training. So overextended was the assigned battlefront position that General Selden realized this regiment would also be needed to man the line. He quickly alerted the 5th Marines commanding officer, Colonel Thomas A. Culhane, Jr., to deploy his regiment, then en route to western Korea, to take over a section of the JAMESTOWN front line instead of assuming reserve positions at Kimpo as originally assigned. General Selden believed that putting another regiment on the main line was essential to carrying out the division's mission, to aggressively *defend* JAMESTOWN Line, not merely to *delay* a Communist advance.

Only a few hours after the 5th Marines had begun its trans-Korea move, helicopters picked up Colonel Culhane, his battalion commanders, and key regimental staff officers and flew them to the relocated division command post (CP) in the west. Here, on 26 March, the regimental commander officially received the change in the 5th Marines mission. Following this briefing, 5th Marines officers reconnoitered the newly assigned area[15] while awaiting the arrival of their units. When the regiment arrived on the 28th, plans had been completed for it to relieve a part of the thinly-held 1st Marines line. On 29 March, the 5th Marines took over the center regimental sector while the 1st Marines, on the right regimental flank, compressed its ranks for a more solid defense.

Frontline units, from the west, were the 1st KMCs, the 5th, and 1st Marines. To the rear, the 7th Marines, designated as division reserve, together with organic and attached units of the division, had established an extensive support and supply area. As a temporary measure, a battalion of the division reserve, 2/7, was detached for defense of the Kimpo Peninsula pending a reorganization of forces in this area. Major logistical facilities were the division airhead, located at K-16 airfield, just southwest of Seoul, and the railhead at Munsan-ni, 25 miles northwest of the capital city and about five miles to the rear of the division sector at its nearest point. Forward of the 1st Marine Division line, outposts were established to enhance the division's security. In the rear area the support facilities, secondary defense lines, and unit command posts kept pace with development of defensive installations on the MLR. Throughout the 1st

[15] Col Thomas A. Culhane, Jr. ltr to Hd, HistBr, G–3 Div, HQMC, dtd 16 Sep 59, hereafter *Culhane ltr*.

Marine Division sector outpost security, field fortifications, and the ground defense net were thorough and intended to deny the enemy access to Seoul.

The Marines' Home in Western Korea[16]

In western Korea, the home of the 1st Marine Division lay in a particularly significant area. (See Map 2.) Within the Marine boundaries ran the route that invaders through the ages had used in their drive south to Seoul. It was the 1st Marine Division's mission to block any such future attempts. One of the reasons for moving the Marines to the west[17] was that the terrain there had to be held at all costs; land in the east, mountainous and less valuable, could better be sacrificed if a partial withdrawal in Korea became necessary. At the end of March 1952, the division main line of resistance stretched across difficult terrain for more than 30 miles, from Kimpo to the British Commonwealth sector on the east, a frontage far in excess of the textbook concept.

Although Seoul was not actually within the area of Marine Corps responsibility, the capital city was only 33 air miles south of the right limiting point of the division MLR and 26 miles southeast of the left. The port of Inchon lay but 19 air miles south of the western end of the division sector. Kaesong, the original site of the truce negotiations, was 13 miles northwest of the nearest part of the 1st Marine Division frontline while Panmunjom was less than 5 miles away and within the area of Marine forward outpost security. From

[16] Unless otherwise noted, the material in this section is derived from: 1stMarDiv ComdD, Mar 52; CIA, *NIS 41B,* South Korea, Chap I, Brief, Section 21, Military Geographic Regions, Section 24, Topography (Washington: 1957–1962); Map, Korea, 1:50,000, AMS Series L 751, Sheets 6526 I and IV, 6527 I, II, III, and IV, 6528 II and III, 6627 III and IV, and 6628 III (prepared by the Engineer, HQ, AFFE, and AFFE/8A, 1952–1954).

[17] The two other reasons were the weakness of the Kimpo defenses and abandonment of plans for an amphibious strike along the east coast. Montross, Kuokka, and Hicks, *USMC Ops Korea,* v. IV, p. 253. Planning for a Marine-led assault had been directed by the EUSAK commander, General Van Fleet, early in 1952. The Marine division CG, General Selden, had given the task to his intelligence and operations deputies, Colonel James H. Tinsley and Lieutenant Colonel Gordon D. Gayle. On 12 March General Van Fleet came to the Marine Division CP for a briefing on the proposed amphibious assault. At the conclusion of the meeting the EUSAK commander revealed his concern for a possible enemy attack down the Korean west coast and told the Marine commander to prepare, in utmost secrecy, to move his division to the west coast. Lynn Montross, draft MS.

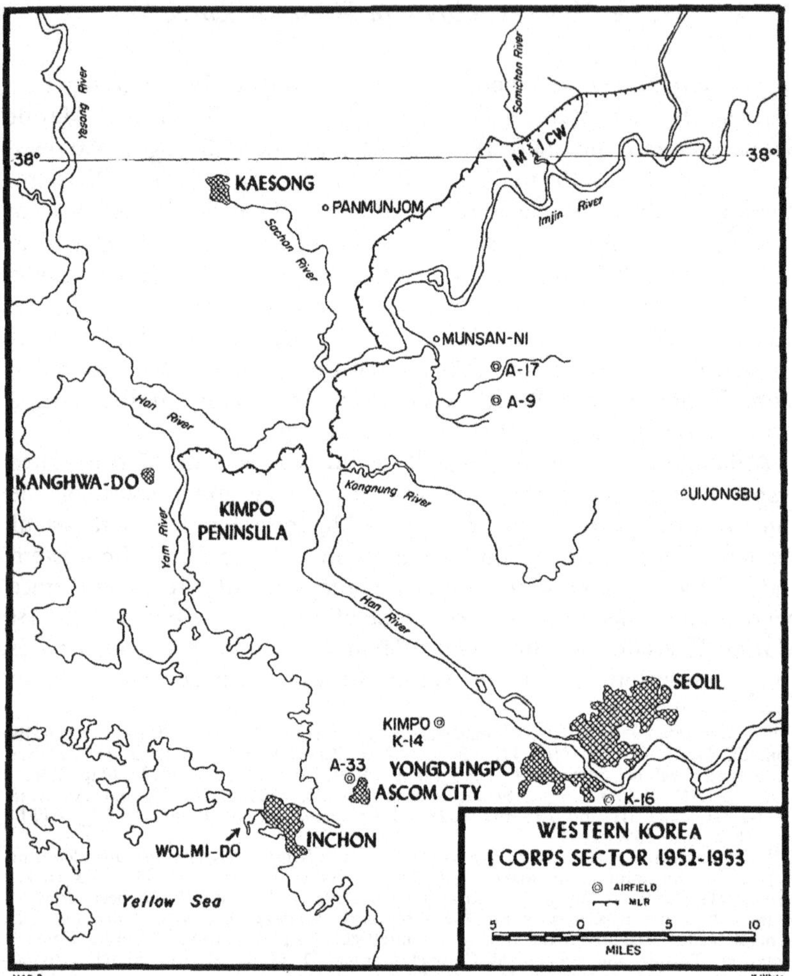
WESTERN KOREA
I CORPS SECTOR 1952-1953
AIRFIELD
MLR
5 0 5 10
MILES
KAESONG
PANMUNJOM
MUNSAN-NI
A-17
A-9
KANGHWA-DO
KIMPO
PENINSULA
UIJONGBU
SEOUL
KIMPO
K-14
A-33
YONGDUNGPO
ASCOM CITY
K-16
WOLMI-DO
INCHON
Yellow Sea
Yesong River
Samichon River
Imjin River
Sachon River
Han River
Kongnung River
Yom River
1 M
1 CW
38°
38°

MAP 2

K. White

the far northeastern end of the JAMESTOWN Line, which roughly paralleled the Imjin River, distances were correspondingly lengthened: Inchon, thus being 39 miles southwest and Kaesong, about 17 miles west.

The area to which the Marines had moved was situated in the western coastal lowlands and highlands area of northwestern South Korea. On the left flank, the division MLR hooked around the northwest tip of the Kimpo Peninsula, moved east across the high ground overlooking the Han River, and bent around the northeast cap of the peninsula. At a point opposite the mouth of the Kongnung River, the MLR traversed the Han to the mainland, proceeding north alongside that river to its confluence with the Imjin. Crossing north over the Imjin, JAMESTOWN followed the high ground on the east bank of the Sachon River for nearly two miles to where the river valley widened. There the MLR turned abruptly to the northeast and generally pursued that direction to the end of the Marine sector, meandering frequently, however, to take advantage of key terrain. Approximately 2½ miles west of the 1st Commonwealth Division boundary, the JAMESTOWN Line intersected the 38th Parallel near the tiny village of Madam-ni.

Within the Marine division sector to the north of Seoul lay the junction of two major rivers, the Imjin and the Han, and a portion of the broad fertile valley fed by the latter. Flowing into the division area from the east, the Imjin River snaked its way southwestward to the rear of JAMESTOWN. At the northeastern tip of the Kimpo Peninsula, the Imjin joined the Han. The latter there changed its course from south to west, flowed past Kimpo and neighboring Kanghwa-do Island, and emptied eventually into the Yellow Sea. At the far western end of the division sector the Yom River formed a natural boundary, separating Kanghwa and Kimpo, as it ran into the Han River and south to the Yellow Sea. To the east, the Sachon River streamed into the Imjin, while the Kongnung emptied into the Han where the MLR crossed from the mainland to Kimpo.

In addition, two north-south oriented rivers flanked enemy positions opposite the Marines and emptied into major rivers in the Marine sector. Northwest of Kimpo, the Yesong River ran south to the Han; far to the northeast, just beyond the March 1952 division right boundary, the Samichon River flowed into the Imjin.

Although the rivers in the Marine division were navigable, they were little used for supply or transportation. The railroads, too, were considered secondary ways, for there was but one line, which ran north out of Seoul to Munsan-ni and then continued towards Kaesong. Below the division railhead, located at Munsan-ni, a spur cut off to Ascom City. Roads, the chief means of surface transport, were numerous but lacked sufficient width and durability for supporting heavy military traffic. Within the sector occupied by the Marines, the main route generally paralleled the railroad. Most of the existing roads south of JAMESTOWN eventually found their way to the logsitic center at Munsan-ni. Immediately across the Imjin, the road net was more dense but not of any better construction.

From the logistical point of view, the Imjin River was a critical factor. Spanning it and connecting the division forward and rear support areas in March 1952 were only three bridges, which were vulnerable to river flooding conditions and possible enemy attack. Besides intersecting the Marine sector, the Imjin formed a barrier to the rear of much of the division MLR, thereby increasing the difficulty of normal defense and resupply operations.

When the Marines moved to the west, the winter was just ending. It had begun in November and was characterized by frequent light snowfalls but otherwise generally clear skies. Snow and wind storms seldom occurred in western Korea. From November to March the mean daily minimum Fahrenheit readings ranged from 15° to 30° above zero. The mean daily maximums during the summer were between the upper 70s and mid-80s. Extensive cloud cover, fog, and heavy rains were frequent during the summer season. Hot weather periods were also characterized by occasional severe winds. Spring and fall were moderate transitional seasons.

Steep-sided hills and mountains, which sloped abruptly into narrow valleys pierced by many of the rivers and larger streams, predominated the terrain in the I Corps sector where the Marines located. The most rugged terrain was to the rear of the JAMESTOWN Line; six miles northeast of the Munsan-ni railhead was a 1,948-foot mountain, far higher than any other elevation on the Marine or Chinese MLR but lower than the rear area peaks supporting the Communist defenses. Ground cover in the division sector consisted of grass, scrub brush, and, occasionally, small trees. Rice fields crowded the valley floors. Mud flats were prevalent in many areas immediately adjacent to the larger rivers which intersected the division territory or virtually paralleled the east and western boundaries of the Marine sector.

The transfer from the Punchbowl in the east to western Korea thus resulted in a distinct change of scene for the Marines, who went from a rugged mountainous area to comparatively level terrain. Instead of facing a line held by predominantly North Korean forces the division was now confronted by the Chinese Communists. The Marines also went from a front that had been characterized by lively patrol action to one that in March 1952 was relatively dormant. With the arrival of the 1st Marine Division, this critical I Corps sector would witness sharply renewed activity and become a focal point of action in the UNC line.

Organization of the 1st Marine Division Area[18]

"To defend" were the key words in the 1st Marine Division mission —"to organize, occupy, and actively defend its sector of Line JAMESTOWN"—in West Korea. General Selden's force to prevent enemy penetration of JAMESTOWN numbered 1,364 Marine officers, 24,846 enlisted Marines, 1,100 naval personnel, and 4,400 Koreans of the attached 1st KMC Regiment. The division also had operational control of several I Corps reinforcing artillery units in its sector. On 31 March, another major infantry unit, the Kimpo Provisional Regiment (KPR) was organized. The division then assumed responsibility for the Kimpo Peninsula defense on the west flank with this Marine-Korean force.

A major reason for transfer of the 1st Marine Division to the west, it will be remembered, had been the weakness of the Kimpo defense. Several units, the 5th KMC Battalion, the Marine 1st Armored Amphibian Battalion, and the 13th ROK Security Battalion (less one company), had been charged with the protection of the peninsula. Their operations, although coordinated by I Corps, were conducted independently. The fixed nature of the Kimpo defenses provided for neither a reserve maneuver element to help repel any

[18] Unless otherwise noted, the material in this section is derived from: *PacFlt EvalRpt* No. 4, Chap. 9; 1stMarDiv, 1stMar, 5thMar, 7thMar, 11thMar ComdDs, Mar 52; 1st KMC RCT Daily Intelligence and Operations Rpts, hereafter KMC Regt UnitRpts, Mar 52; Kimpo ProvRegt ComdDs, hereafter KPR ComdDs, Mar–Apr 52.

enemy action that might develop nor a single commander to coordinate the operations of the defending units.

These weaknesses become more critical in consideration of the type of facilities at Kimpo and their proximity to the South Korean Capital. Seoul lay just east of the base of Kimpo Peninsula, separated from it only by the Han River. Located on Kimpo was the key port of Inchon and two other vital installations, the logistical complex at Ascom City and the Kimpo Airfield (K-14). All of these facilities were indispensable to the United Nations Command.

To improve the security of Kimpo and provide a cohesive, integrated defense line, CG, 1st Marine Division formed the independent commands into the Kimpo Provisional Regiment. Colonel Edward M. Staab, Jr., was named the first KPR commander. His small headquarters functioned in a tactical capacity only without major administrative duties. The detachments that comprised the KPR upon its formation were:

Headquarters and Service Company, with regimental and company headquarters and a communication platoon;
1st Armored Amphibian Battalion, as supporting artillery;
5th KMC Battalion;
13th ROK Security Battalion (–);
One battalion from the reserve regiment of the 1st Marine Division (2/7), as the maneuver element;
Company A, 1st Amphibian Tractor Battalion;
Company B, 1st Shore Party Battalion, as engineers;
Company D, 1st Medical Battalion;
Reconnaissance Company (–), 1st Marine Division;
Detachment, Air and Naval Gunfire Liaison Company (ANGLICO), 1st Signal Battalion;
Detachment, 181st Counterintelligence Corps Unit, USA;
Detachment, 61st Engineer Searchlight Company, USA; and the 163rd Military Intelligence Service Detachment, USA.

The Kimpo Regiment, in addition to maintaining security of the division left flank, was assigned the mission to "protect supporting and communication installations in that sector against airborne or ground attack."[10] Within the division, both the artillery regiment and

[10] KPR ComdD, Mar 52, p. 13.

the motor transport battalion were to be prepared to support tactical operations of Colonel Staab's organization.

For defense purposes, the KPR commander divided the peninsula into three sectors. The northern one was manned by the KMC battalion, which occupied commanding terrain and organized the area for defense. The southern part was defended by the ROK Army battalion, charged specifically with protection of the Kimpo Airfield and containment of any attempted enemy attack from the north. Both forces provided for the security of supply and communication installations within their areas. The western sector, held by the amphibian tractor company, less two platoons, had the mission of screening traffic along the east bank of the Yom River, that flanked the western part of the peninsula. Providing flexibility to the defense plan was the maneuver unit, the battalion assigned from the 1st Marine Division reserve.

The unit adjacent to the KPR[20] in the division line in late March was the 1st Korean Marine Corps Regiment, which had been the first division unit to deploy along JAMESTOWN. The KMC Regiment, command by Colonel Kim Dong Ha,[21] had assumed responsibility for its portion of JAMESTOWN at 0400 on 20 March with orders to organize and defend its sector. The regiment placed two battalions, the 3d and 1st, on the MLR and the 2d in the rear. Holding down the regimental right of the sector was the 1st Battalion, which had shared its eastern boundary with that of Colonel Wade's 1st Marines until 29 March when the 5th Marines was emplaced on the MLR between the 1st KMC and 1st Marines.

The 1st Marines regimental right boundary, which on the MLR was 1,100 yards north of the 38th Parallel, separated the 1st Marine Division area from the western end of the 1st Commonwealth Division, then held by the 25th Canadian Infantry Brigade. In late March, Colonel Wade's 2/1 (Lieutenant Colonel Thell H. Fisher) and 3/1 (Lieutenant Colonel Spencer H. Pratt) manned the frontline positions while 1/1 (Lieutenant Colonel John H. Papurca), less Company A, was in reserve. The regiment was committed to the defense of its part of the division area and improvement of its ground

---

[20] The following month the 1st Amphibian Tractor Battalion would be added to the four regiments on line, making a total of five major units manning the 1stMarDiv front. It was inserted between the Kimpo and 1st KMC regiments.

[21] Commandant, Korean Marine Corps ltr to CMC, dtd 20 Sep 66, hereafter *CKMC ltr.*

positions. In the division center sector Colonel Culhane's 1/5 (Lieutenant Colonel Franklin B. Nihart) and 3/5 (Lieutenant Colonel William S. McLaughlin) manned the left and right battalion MLR positions, with 2/5 (Lieutenant Colonel William H. Cushing) in reserve. The latter unit was to be prepared either to relieve the MLR battalions or for use as a counterattack force.

It did not take the Marines long to discover the existence of serious flaws in the area defense which made it questionable whether the Allied line here could have successfully withstood an enemy attack. While his Marine units were effecting their relief of JAMESTOWN, Colonel Wade noted that "field fortifications were practically nonexistent in some sections."[22] General Selden later pointed out that "populated villages existed between opposing lines. Farmers were cultivating their fields in full view of both forces. Traffic across the river was brisk."[23] A member of the division staff reported that there was "even a school operating in one area ahead of the Marine lines."[24] In addition to these indications of sector weakness, there was still another. Although the ROK division had placed three regiments in the line, when the two Marine regiments relieved them there were then more men on JAMESTOWN due to the greater personnel strength of a Marine regiment. Nevertheless, the division commander was still appalled at the width of the defense sector assigned to so few Marines.

At division level, the reserve mission was filled by Colonel Russell E. Honsowetz', 7th Marines, minus 2/7 (Lieutenant Colonel Noel C. Gregory), which on 30 March became the maneuver force for the Kimpo Regiment. As the division reserve, the regiment was to be prepared to assume at any time either a defensive or offensive mission of any of the frontline regiments. In addition, the reserve regiment was to draw up counterattack plans, protect the division rear, improve secondary line defenses, and conduct training, including tank-infantry coordination, for units in reserve. The 7th Marines, with 3/7 (Lieutenant Colonel Houston Stiff) on the left and 1/7 (Lieutenant Colonel George W. E. Daughtry) on the right, was emplaced in the vicinity of the secondary defense lines, WYOMING and KANSAS, to the rear of the 5th and 1st Marines.

---

22 1stMar ComdD, Mar 52, p. 2.
23 1stMarDiv ComdD, Jun 52, App IX, p. 1.
24 LtCol Harry W. Edwards comments on preliminary draft MS, *ca.* Sep 59.

Another regiment located in the rear area was the 11th Marines. Its artillery battalions had begun displacement on 17 March and completed their move by 25 March. Early on the 26th, the 11th Marines resumed support of the 1st Marine Division. While the Marine artillery had been en route, U.S. Army artillery from I Corps supported the division. With the arrival on the 29th of the administrative rear echelon, the Marine artillery regiment was fully positioned in the west.

For Colonel Frederick P. Henderson, who became the division artillery commander on 27 March, operational problems in western Korea differed somewhat from those experienced in the east by his predecessor, Colonel Bruce T. Hemphill. The most critical difficulty, however, was the same situation that confronted General Selden—the vast amount of ground to be covered and defended, and the insufficient number of units to accomplish this mission. To the artillery, the wide division front resulted in spreading the available fire support dangerously thin. Placement of 11th Marines units to best support the MLR regiments created wide gaps between each artillery battalion, caused communication and resupply difficulties, prevented a maximum massing of fires, and made redeployment difficult.[25]

In making use of all available fire support, the artillery regiment had to guard not only against the duplication of effort in planning or delivery of fires, but also against firing in the Panmunjom peace corridor restricted areas, located near the sector held by the Marine division's center regiment. Moreover, the artillerymen had to maintain a flexibility sufficient to place the weight of available fire support on call into any zone of action.

Other difficulties were more directly associated with the nature of the sector rather than with its broad expanse. The positioning of the division in the west, although close to the coast, put the Marines beyond the range of protective naval gunfire. The sparse and inadequate road net further aggravated the tactical and logistical problems caused by wide separation of units. Finally, the cannoneers had exceptionally heavy demands placed on them due to the restricted amount of close air support allocated to frontline troops under operational procedures employed by Fifth Air Force. This command

---

[25] Col Frederick P. Henderson ltr to Hd, HistBr, G–3 Div, HQMC, dtd 25 Aug 59, hereafter *Henderson ltr I.*

had jurisdiction over the entire Korean air defense system, including Marine squadrons.

Manning the main line of resistance also frequently presented perplexing situations to the infantry. There had been little time for a thorough reconnaissance and selection of positions by any of the frontline regiments. When the 1st Marines moved into its assigned position on the MLR, the troops soon discovered many minefields, "some marked, some poorly marked, and some not marked at all."[26] Uncharted mines caused the regiment to suffer "some casualties the first night of our move and more the second and third days."[27] As it was to turn out, during the first weeks in the I Corps sector, mines of all types caused 50 percent of total Marine casualties.

A heavy drain on the limited manpower of Marine infantry regiments defending JAMESTOWN was caused by the need to occupy an additional position, an outpost line of resistance (OPLR). This defensive line to the front of the Marine MLR provided additional security against the enemy, but decreased the strength of the regimental reserve battalion, which furnished the OPLR troops. The outposts manned by the Marines consisted of a series of strongpoints built largely around commanding terrain features that screened the 1st Marine Division area. The OPLR across the division front was, on the average, about 2,500 yards forward of the MLR. (See Map 3.)

To the rear of the main line were two secondary defensive lines, WYOMING and KANSAS. Both had been established before the Marines arrived and both required considerable work, primarily construction of bunkers and weapons emplacements, to meet General Selden's strict requirement for a strong defensive sector. Work in improving the lines, exercises in rapid battalion tactical deployment by helicopter, and actual manning of the lines were among the many tasks assigned to the division reserve regiment.

Rear and frontline units alike found that new regulations affected combat operations with the enemy in West Korea. These restrictions were a result of the truce talks that had taken place first at Kaesong and, later, at Panmunjom. In line with agreements reached in October 1951:

> Panmunjom was designated as the center of a circular neutral zone of a 1,000 yard radius, and a three mile radius around Munsan and Kaesong was

[26] Col Sidney S. Wade ltr to Deputy AsstCofS, G–3, HQMC, dtd 25 Aug 59.
[27] *Ibid.*

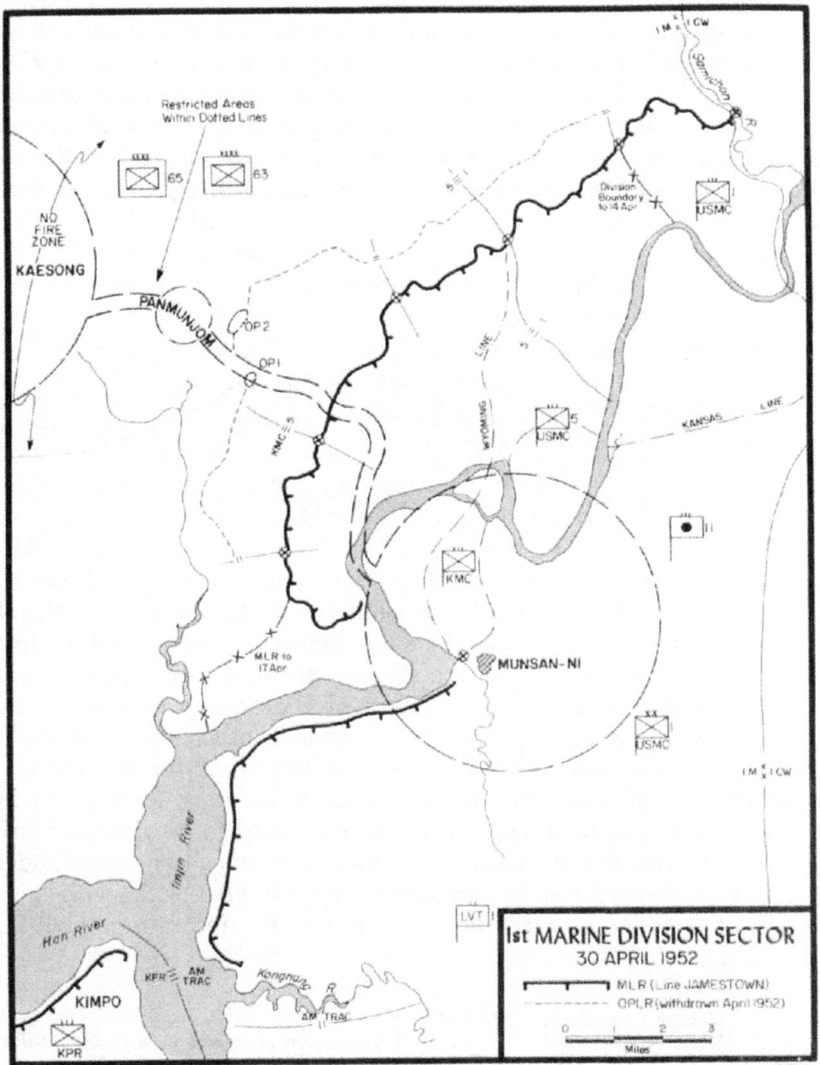

Restricted Areas Within Dotted Lines
NO FIRE ZONE
KAESONG
PANMUNJOM
OP 2
OP 1
65
63
KMC
Division Boundary to 14 Apr
USMC
Samichon R
WYOMING LINE
KANSAS LINE
5 USMC
KMC
MUNSAN-NI
MLR to 17 Apr
USMC
Imjin River
Han River
KIMPO
KPR
AM TRAC
Kongnung R
LVT
1st MARINE DIVISION SECTOR
30 APRIL 1952
MLR (Line JAMESTOWN)
OPLR (withdrawn April 1952)
Miles

MAP 3

K White

also neutralized, as well as two hundred meters on either side of the Kaesong–Munsan road.[28]

To prevent the occurrence of any hostile act within this sanctuary, Lieutenant General John W. O'Daniel, I Corps commander, ordered that an additional area, forward of the OPLR, be set aside. This megaphone-shaped zone "could not be fired into, out of, or over."[29] It was adjacent to the OPLR in the division center regimental sector, near its left boundary, and took a generally northwest course. Marines reported that the Communists knew of this restricted zone and frequently used it for assembly areas and artillery emplacements.

The 1st Marine Aircraft Wing[30]

When the 1st Marine Division moved to western Korea in March 1952, the two 1st Marine Aircraft Wing units that had been in direct support of the ground Marines also relocated. Marine Observation Squadron 6 (VMO–6) and Marine Helicopter Transport Squadron 161 (HMR–161) completed their displacements by 24 March from their eastern airfield (X–83) to sites in the vicinity of the new division CP. HMR–161, headed by Colonel Keith B. McCutcheon, set up headquarters at A–17,[31] on a hillside 3½ miles southeast of Munsan-ni, the division railhead, "using a couple of rice paddies as our L. Z. (Landing Zone)."[32] The squadron rear echelon, including the machine shops, was maintained at A–33, near Ascom City. About 2½ miles south of the helicopter forward site was an old landing strip, A–9, which Lieutenant Colonel William T. Herring's observation squadron used as home field for its fixed and rotary wing aircraft. (For location of 1st MAW units see Map 4.) In West Korea, VMO–6 and HMR–161 continued to provide air transport for tactical and logistical missions. Both squadrons were under operational control of the division, but administered by the wing.

Commanding General of the 1st MAW, since 27 July 1951,

---

[28] Rees, *Korea,* p. 295.

[29] 1stMarDiv ComdD, Mar 52, p. 7.

[30] Unless otherwise noted, the material in this section is derived from *PacFlt EvalRpt* No. 4, Chap. 10; 1stMarDiv ComdD, Mar 52; 1st MAW ComdDs, Mar–Apr 52.

[31] In Korea, fields near U.S. Army installations were known as "A"; major airfields carried a "K" designation; and auxiliary strips were the "X" category.

[32] MajGen Keith B. McCutcheon comments on draft MS, dtd 1 Sep 66.

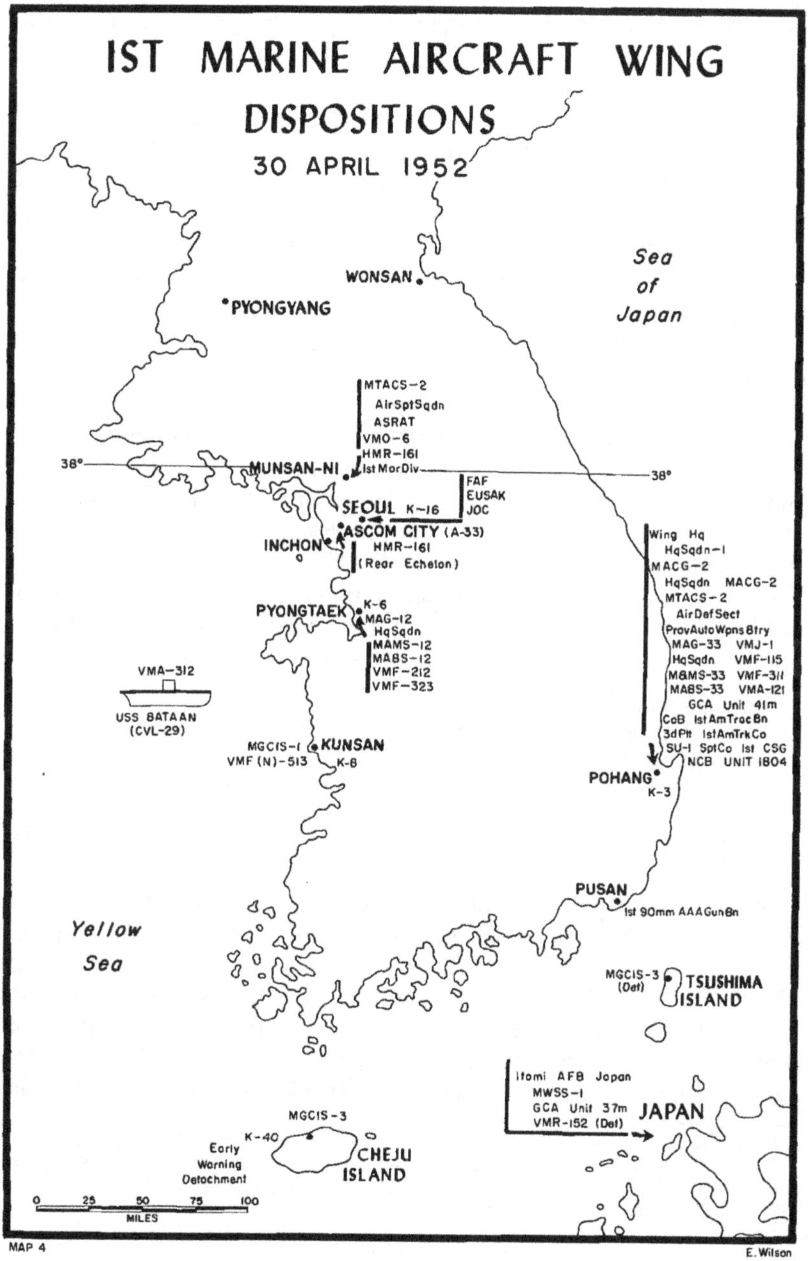

MAP 4

E. Wilson

was Major General Christian F. Schilt,[33] a Marine airman who had brought to Korea a vast amount of experience as a flying officer. Entering the Marine Corps in June 1917, he had served as an enlisted man with the 1st Marine Aeronautical Company in the Azores during World War I. Commissioned in 1919, he served in a variety of training and overseas naval air assignments. As a first lieutenant in Nicaragua, he had been awarded the Medal of Honor in 1928 for his bravery and "almost superhuman skill" in flying out Marines wounded at Quilali.[34] During World War II, General Schilt had served as 1st MAW Assistant Chief of Staff, at Guadalcanal, was later CO of Marine Aircraft Group 11, and participated in the consolidation of the Southern Solomons and air defense of Peleliu and Okinawa.

As in past months, the majority of General Schilt's Marine aircraft in Korea during March 1952 continued to be under operational control of Fifth Air Force. In turn, FAF was the largest subordinate command of Far East Air Forces (FEAF), headquartered at Tokyo. The latter was the U.S. Air Force component of the Far East Command and encompassed all USAF installations in the Far East. The FAF–EUSAK Joint Operations Center (JOC) at Seoul coordinated and controlled all Allied air operations in Korea. Marine fighter and attack squadrons were employed by FAF to:

> Maintain air superiority.
> Furnish close support for ground units.
> Provide escort [for attack aircraft].
> Conduct day and night reconnaissance and fulfill requests.
> Effect the complete interdiction of North Korean and Chinese Communist forces and other military targets that have an immediate effect upon the current tactical situation.[35]

Squadrons carrying out these assignments were attached to Marine Aircraft Groups (MAGs) 12 and 33. Commanded by Colonel Luther S. Moore, MAG–12 and its two day attack squadrons (VMF–212 and VMF–323) in March 1952 was still located in eastern Korea (K–18, Kangnung). The Marine night-fighters of VMF(N)–513 were also here as part of the MAG–12 group. Farther removed from the immediate battlefront was Colonel Martin

[33] DivInfo, HQMC, Biography of General Christian F. Schilt, USMC (Ret.), Jun 59 rev.
[34] Robert Sherrod, *History of Marine Corps Aviation in World War II* (Washington: Combat Forces Press, 1952), p. 26, hereafter Sherrod, *Marine Aviation.*
[35] 1st MAW ComdD, Mar 52, p. 2.

A. Severson's MAG–33, located at K–3 (Pohang), with its two powerful jet fighter squadrons (VMFs–115 and –311) and an attack squadron (VMA–121). A new MAG–33 unit was Marine Photographic Squadron 1 (VMJ–1), just formed in February 1952 and commanded by Major Robert R. Read.

In addition to its land-based squadrons, one 1st MAW unit was assigned to Commander, West Coast Blockading and Patrol Group, designated Commander, Task Group 95.1 (CTG 95.1). He in turn assigned this Marine unit to Commander, Task Element 95.11 (CTE 95.11), whose ships comprised the West Coast Carrier Element. Marine Attack Squadron 312 (VMA–312) was at this time assigned to CTE 95.11. In late March squadron aircraft were based on the escort carrier USS *Bairoko* but transferred on 21 April to the light carrier *Bataan*.[36] Operating normally with a complement of 21 F4U–4 propeller-driven Corsair aircraft, VMA–312 had the following missions:

> To conduct armed air reconnaissance of the West Coast of Korea from the United Nations front lines northward to latitude 39°/15′ N.
>
> Attack enemy shipping and destroy mines.
>
> Maintain surveillance of enemy airfields in the Haeju–Chinnampo region.[37]
>
> Provide air spot services to naval units on request.
>
> Provide close air support and armed air reconnaissance services as requested by Joint Operations Center, Korea (JOC KOREA).
>
> Conduct air strikes against coastal and inland targets of opportunity at discretion.
>
> Be prepared to provide combat air patrol to friendly naval forces operating off the West Coast of Korea.
>
> Render SAR [search and rescue] assistance.

Because they were under operational control of Fifth Air Force, 1st MAW flying squadrons, except those assigned to CTG 95.1 and 1st Marine Division control, did not change their dispositions in March. Plans were under way at this time, however, to relocate one of the aircraft groups, MAG–12, to the west.

On 30 March the ground element of the night-fighters redeployed

---

[36] Unit commanders also changed about this time. Lieutenant Colonel Robert E. Smith, Jr. assumed command of the Checkerboard squadron from Lieutenant Colonel Joe H. McGlothlin, on 9 April.

[37] *PacFlt EvalRpt* No. 4, p. 10–75. The Haeju–Chinnampo region, noted in the surveillance mission, is a coastal area in southwestern North Korea between the 38th and 39th Parallels.

from its east coast home field to K–8 (Kunsan), on the west coast, 105 miles south of Seoul. Lieutenant Colonel John R. Burnett's VMF (N)–513 completed this relocation by 11 April without loss of a single day of flight operations. On 20 April the rest of MAG–12,[38] newly commanded since the first of the month by Colonel Elmer T. Dorsey, moved to K–6 (Pyongtaek), located 30 miles directly south of the South Korean capital.

Marine aircraft support units were also located at K–3 and at Itami Air Force Base, on Honshu, Japan. Under direct 1st MAW control were four ground-type logistical support units with MAG–33, a Provisional Automatic Weapons Battery from Marine Air Control Group 2 (MACG–2), and most of wing headquarters. This last unit, commanded by Colonel Frederick R. Payne, Jr., included the 1st 90mm AAA Gun Battalion (based at Pusan and led by Colonel Max C. Chapman), and a detachment of Marine Transport Squadron 152 (VMR–152), which had seven Douglas four-engine R5D transports. This element and the wing service squadron were based at Itami.

Marines, and others flying in western Korea, found themselves restricted much as Marines on the ground were. One limitation resulted from a FAF–EUSAK agreement in November 1951 limiting the number of daily close air support sorties across the entire Eighth Army line. This policy had restricted air activity along the 155-mile Korean front to 96 sorties per day. The curtailment seriously interfered with the Marine type of close air support teamwork evolved during World War II, and its execution had an adverse effect on Marine ground operations as well. A second restriction, also detrimental to Marine division and wing efficiency, was the prohibitive cushion Fifth Air Force had placed around the United Nations peace corridor area north of the Marine MLR. This buffer no-fly, no-fire zone which had been added to prevent violation of the UN sanctuary by stray hits did not apply, of course, to the Communists.

---

[38] VMFs–212 (LtCol Roert L. Bryson) and –323 (LtCol Richard L. Blume) left an east coast field for a flight mission over North Korea and landed at K–6 thereafter, also completing the move without closing down combat operations. The relocation in airfields was designed to keep several squadrons of support aircraft close to the 1st Marine Division. Col E. T. Dorsey ltr to Hd, HistBr, G–3 Div, HQMC, dtd 7 Sep 66.

The Enemy[39]

Directly beyond the 1st Marine Division sector, to the west and north, were two first-rate units of the Chinese Communist Forces, the 65th and 63d CCF Armies. Together, they totaled approximately 49,800 troops in late March 1952. Opposite the west and center of the Marine division front was the 65th CCF Army, with elements of the 193d Division across from the KPR and the 194th Division holding positions opposing the KMC regiment. Across from the Marine line in the center was the 195th Division of the 65th CCF Army, which had placed two regiments forward. North of the division right sector lay the 188th Division, 63d CCF Army, also with two regiments forward. The estimated 15 infantry battalions facing the Marine division were supported by 10 organic artillery battalions, numbering 106 guns, and varying in caliber from 75 to 155mm.[40] In addition, intelligence reported that the 1st CCF Armored Division and an unidentified airborne brigade were located near enough to aid enemy operations.

Chinese infantry units were not only solidly entrenched across their front line opposite the Marine division but were also in depth. Their successive defensive lines, protected by minefields, wire, and other obstacles, were supported by artillery and had been, as a result of activities in recent months, supplied sufficiently to conduct continuous operations. Not only were enemy ground units well-supplied, but their CCF soldiers were well disciplined and well led. Their morale was officially evaluated as ranging from good to excellent. In all, the CCF was a determined adversary of considerable ability, with their greatest strength being in plentiful combat manpower.

Air opposition to Marine pilots in Korea was of unknown quantity and only on occasion did the caliber of enemy pilots approach that of the Americans. Pilots reported that their Chinese counterparts generally lacked overall combat proficiency, but that at times their "aggressiveness, sheer weight of numbers, and utter disregard for losses have counterbalanced any apparent deficiencies."[41] The

---

[39] Unless otherwise noted, the material in this section is derived from: *PacFlt EvalRpt* No. 4, Chaps. 9, 10; 1stMarDiv ComdD, Mar 52.

[40] The Korean Marine Corps placed the artillery count at 240 weapons ranging from 57 to 122mm. *CKMC ltr.*

[41] *PacFlt EvalRpt*, No. 4, p. 10–38.

Communists had built their offensive potential around the Russian MIG–15 jet fighter-interceptor. Use of this aircraft for ground support or ground attack was believed to be in the training stage only. The Chinese had also based their air defense on the same MIG plus various types of ground antiaircraft (AA) weapons, particularly the mobile 37mm automatic weapons and machine guns that protected their main supply routes. In use of these ground AA weapons, enemy forces north of the 38th Parallel had become most proficient. Their defense system against UNC planes had been steadily built up and improved since stabilization of the battle lines in 1951, and by March 1952 was reaching a formidable state.

As the more favorable weather for ground combat approached toward the end of March, the CCF was well prepared to continue and expand its operations. Enemy soldiers were considered able to defend their sector easily with infantry and support units. Division intelligence also reported that Chinese ground troops had the capability for launching limited objective attacks to improve their observation of Marine MLR rear areas.

Initial CCF Attack[42]

Whether by intent or default, the Chinese infantry occupying the enemy forward positions did not interfere with the Marine relief. With assumption of sector responsibility by the division early on 25 March, the initial enemy contact came from Chinese supporting weapons. Later that day the two division frontline regiments, the 1st and 5th Marines, received 189 mortar and artillery shells in their sectors which wounded 10 Marines. One man in the 1st Marines was killed by sniper fire on 25 March; in the same regiment, another Marine was fatally wounded the following day. Forward of the lines, the day after the division took over, there was no ground action by either side.

During the rest of the month, the tempo of activities on both sides increased. Marines began regular patrol actions to probe and ambush the enemy. Division artillery increased its number of observed missions by the end of the month. By this time the CCF had also begun

[42]Unless otherwise noted, the material in this section is derived from: 1stMarDiv ComdDs, Mar–Apr 52; KMC Regt UnitRpt 31, dtd 2 Apr 52.

to probe the lines of the Marine regimental sectors. In these ground actions to reconnoiter and test division defenses, the Chinese became increasingly bold, with the most activity on 28 March. Between 25–31 March, the first week on JAMESTOWN, some 100 Chinese engaged in 5 different probing actions. Most of these were against the 1st KMC Regiment on the left flank of the division MLR.

It was no wonder that the Chinese concentrated their effort against the Korean Marines, for they held the area containing Freedom Gate, the best of the three bridges spanning the Imjin. Both of the other two, X-Ray and Widgeon, were further east in the division sector. If the enemy could exploit a weak point in the KMC lines, he could attack in strength, capture the bridge, and turn the division left flank, after which he would have a direct route to Seoul.[43] Without the bridge in the KMC sector, the division would be hard pressed, even with helicopter lift, to maneuver or maintain the regiments north of the Imjin.

On 1 April, at about 2130, the CCF began pounding the front-line companies in the KMC area with an artillery preparation. A half hour later, the enemy attacked an outpost and the main line. First to engage the Chinese were the OPLR troops of the KMC 1st Company, 1st Battalion, on the regimental right. There, a Chinese company forced an opening between friendly outposts and reached a point about 200 yards short of the MLR and just north of a road leading to the main bridge over the Imjin. While this attack was in progress, another CCF company hit the outpost line further south. This attack, less successful, ended far short of the MLR and about a half-mile south of the bridge road. Both enemy companies withdrew at about 2345.

To the left of the 1st Battalion, the 3d was receiving the brunt of this initial CCF attack. The 9th, 11th, and 10th Companies (deployed in that order from west to east, in the left battalion sector), had been engaged by the same preliminary 30-minute shelling. At 2200, when four CCF squads attacked the two companies on the left, an enemy company hit the left end of the 10th Company, occupied by the 2d Platoon. About midnight the South Koreans, under fire from both flanks and under heavy frontal assault, were forced to withdraw. In the rear, the company commander pulled the 1st

---

[43] *Henderson ltr I.*

Platoon from the line, ordered the 3d to extend left to cover both sectors, and led a counterattack with the 1st Platoon and elements of the 2d. Positions were quickly restored by the KMC action.

Soon after it had hurled the Chinese back across the OPLR, the 1st Battalion was subjected to a second attack. An enemy unit, estimated to be a company, engaged a 1st Company platoon briefly. When the KMCs returned heavy defensive fires, the Communists pulled back but struck again at 0300. After a 20-minute fire fight, the Chinese company retreated.

This action on 1–2 April cost the attackers 2 killed, 34 estimated killed, and 10 estimated wounded. For the KMC, casualties were 2 killed, 10 wounded. To all 1st Division Marines, the successful defense by the 1st KMC regimental Marines was heartening. It had preserved not only the division western flank but also the vital link over the Imjin.

Subsequent CCF Attacks[44]

Following his attempted assault against the KMC regiment, the enemy opposite the 1st Marine Division reverted to a passive defense. Except for a probe late on 2 April of the far eastern line held by Lieutenant Colonel Pratt's 3/1 and two patrols that scouted MLR positions in the western Korean Marine area that same date, Communist offensive measures consisted largely of artillery and mortar fire. Chinese line units appeared to concentrate on improving their dugouts and trench systems. Marines reported frequent sightings of enemy groups working in and around their forward trenches.

Marine division troops, too, were busy fortifying their defensive positions. On the Kimpo Peninsula they dug gun emplacements and erected camp facilities for the newly activated Kimpo Provisional Regiment. North of the Han, mine clearance and construction of trenchworks and fortifications was the order of the day for most Marines. Other Marines patrolled forward of the lines as a major aspect of the division's continuous active defense. During daylight hours, MLR regiments dispatched reconnaissance and combat patrols and sent out snipers, armed with telescope-equipped M–1 rifles.

---

[44] Unless otherwise noted, the material for this section is derived from: 1stMarDiv, 1stMar, 5thMar, KPR ComdDs, Apr 52; KMC Regt UnitRpt 35, dtd 16 Apr 52.

Division tanks firing from temporary gun slots on the main line and artillery batteries emplaced in rear area dugouts hammered away at enemy positions and disposed of his patrols. At night, harassing and interdicting (H&I) artillery fires and infantry raids continued to keep the Communists off-balance.

A combat raid on 5 April typified the extensive Marine division night activities forward of the line. Conducted by three platoons, less a squad, of the KMC 10th Company, the raiding party had the mission of capturing prisoners. Departing the MLR at 2300, the Korean Marines worked their way over the low ground and then crossed the Sachon River. Immediately thereafter the raid leader, who was the 10th Company commander (First Lieutenant No Won Keun) dispatched two squad-sized ambushes along the patrol route. The raiders then continued northwest toward their objective, an area near the village of Tonggang-ni, a half mile beyond the river. When about 50 yards from its objective, the patrol ran into tactical wire and an enemy sentry, who alerted his unit by rifle fire. The KMC raiders opened up and called in pre-planned mortar and artillery support. The CCF defenders replied immediately with rifles and machine gun fire.

To complete the maneuver, the patrol leader positioned his machine guns to fire on the Communist flanks and directed one platoon to prepare for a frontal assault on the defenders. At 0148, the 1st Platoon attacked from the right. A minute later the 2d Platoon charged headlong at the defenders. Hand-to-hand fighting followed until the Chinese broke contact and disappeared into bunkers within the trenchline. From inside, the CCF soldiers continued the battle, firing through gun revetments and wounding several KMC pursuers in the legs. After 30 minutes had passed, the South Korean assault troops observed enemy reinforcements moving in from the northwest. At 0230, the Marine patrol withdrew under the cover of artillery, reaching its battalion MLR at 0400. The raiders brought back seven civilians found in the area and several Russian-made carbines. At the cost of 2 killed and 18 wounded, the KMCs inflicted casualties totaling 12 counted killed and 25 estimated wounded.

Other division patrols similarly took into custody civilians living between the MLR and OPLR. It was also the job of these patrols to destroy buildings that the enemy had used. On the night of 5 April, 5th Marines patrols apprehended 34 civilians, and a wounded enemy soldier. The day before, a patrol from 2/1 had also captured a Chinese soldier.

On 12 and 13 April, the enemy stepped up his ground actions. He launched two probes against the 5th Marines occupying the center regimental sector. Both attempts were beaten back. The 1st Marines on the extreme right flank encountered little hostile activity, but in the western KMC sector, Chinese shelling increased noticeably. The following day the artillery picked up again, accompanied by several infantry probes directed against the two KMC frontline battalions. To the right, the Chinese also tested 5th Marines lines again. On the far right, in the area held by the 1st Marines, an air alert was sounded from 0410 to 0726, but no enemy aircraft appeared. By mid-month, the Chinese were dispatching fewer infantry probes but firing a greater number of artillery and mortar shells toward the division line. The enemy even sent 25 rounds to Kimpo, where a total of only 4 had fallen during the first two weeks in April.

Ushering in the second half of April was another Communist attack, this one on 15–16 April and to be the last that month against the central part of the Marine Division sector. This attempt to breach the Marine lines was directed against Company E of 2/5, manning an outpost position on the OPLR. The rest of the battalion was now holding the left sector of the center regimental front, having assumed its new mission on line three days earlier in relief of 1/5, which reverted to the role of regimental reserve. Northwest of the 5th Marines MLR, the Company E commander, Captain Charles C. Matthews, had placed a reinforced rifle platoon. His Marines had occupied several dug-in positions near the top of a 400-foot hill, known as Outpost 3 (OP 3). (See Map 5.) The platoon had been improving this outpost area and fortifications so that the bunkers could be employed for living and fighting.[45] During the afternoon and again at dusk on 15 April the Communists had shelled this location. One Marine was wounded in the second firing.

At 2330 on 15 April, Company E reported that a green flare cluster had just burst over Hill 67, approximately 1,900 yards southwest of OP 3 and just beyond the OPLR. This signal triggered a 20-minute heavy enemy preparation of 76mm artillery and 120mm mortars on the friendly outpost and its supporting mortar position. Ten minutes

---

[45] Chapter III discusses in detail the construction of bunkers.

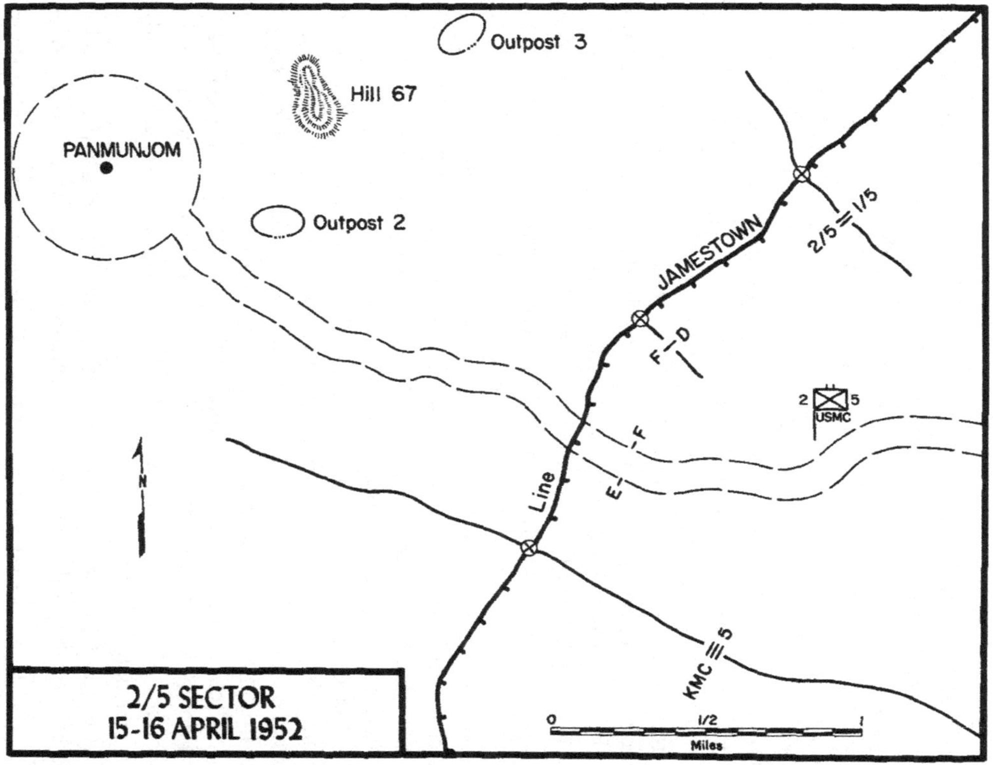
Outpost 3
Hill 67
PANMUNJOM
Outpost 2
JAMESTOWN
Line
2/5 = 1/5
F — D
2 5
USMC
F
E
N
KMC = 5
0
1/2
1
Miles
2/5 SECTOR
15-16 APRIL 1952
MAP 5

K. WHITE

before midnight, another green flare exploded over the same height, and the shelling stopped. After five minutes the signal reappeared. Immediately thereafter, the Chinese shifted their artillery and mortar fire to an area west of the OP 3 mortar site and north of a Company F observation post. At the same time, the enemy attacked Outpost 3.

Initially, the Chinese struck the Marine defenses in a frontal assault, but as the fighting progressed enemy forces quickly enveloped the outpost and charged it simultaneously from three sides. The vastly outnumbered Marine defenders withdrew into a tight perimeter at the southeastern corner of the outpost where their defending firepower prevented the enemy from seizing the position. Within 15 minutes the enemy had surrounded the Marines and severed the outpost communications, but could not take the outpost. The CCF soldiers then pulled back and let their artillery soften OP 3 while they regrouped for another assault. The Chinese soon stormed the outpost a second time, but were again unsuccessful. Moreover, they lost three of their men who were captured by the tenacious 2/5 defenders.

The fighting continued until 0315, reaching a hand-to-hand clash at one stage. In addition to mortar and artillery fire, the enemy employed small arms, automatic weapons, hand and stick-type grenades, bangalore torpedoes, and 57mm recoilless rifles. During the attack, patrols were sent out from the MLR and OP 2, to the west, to reestablish contact and help with casualty evacuation.

Well to the rear of the outpost and unknown to its occupants, intelligence personnel intercepted a Chinese message ordering the Communists to withdraw. Immediately, friendly artillery fired on all known escape routes available to the attackers. Despite this interdicting fire, the enemy soldiers managed to withdraw without further loss. Their unsuccessful thrust against the 2/5 OPLR cost the Chinese 25 known killed, 25 estimated killed, 45 known wounded, and 3 prisoners. Marine casualties were 6 killed, 5 missing, and 25 wounded and evacuated.[46]

Why the Chinese had selected OP 3 for their mid-April attack is not known. Several theories, however, have been advanced by those

[46] One of those wounded was Corporal Duane E. Dewey, a machine gunner. He was wounded twice, in fact, the second time from an exploding enemy grenade which he had rolled upon to shield two nearby comrades. Dewey somehow survived, and the following March, after release from the Marine Corps, he went to the White House where he received the Medal of Honor, the first to be presented by the new President, Dwight D. Eisenhower. (Duane E. Dewey Biog. File)

involved in the action. Colonel Culhane, the regimental commander, believed that the enemy incursion "was the direct result of the aggressive patrols that frequently used the outpost as a point of departure. . . . "[47] Brigadier General Merrill B. Twining, the assistant division commander since 22 March, declared that the position was too large for a reinforced platoon to hold.[48] Perhaps the Chinese had harbored the same thoughts before the night of 15–16 April.

Just before its OPLR was withdrawn in favor of an observation line, the 1st Korean Regiment was struck by the Chinese in the area immediately north of the 1–2 April clash. Beginning at 0100 on 17 April, the enemy placed a 15-minute preparatory fire on the left flank of the 3d Battalion, occupying the regimental right sector. The CCF then probed friendly lines in and around the area pounded during the preliminary fires. Three separate attacks took place before 0400, when the Communists withdrew. In these probes, the Chinese made free use of automatic weapons; the enemy's well-coordinated action attested to their training and discipline. Confirmed casualties were 36 CCF and 2 Koreans killed. The KMCs suffered 5 wounded and estimated that 70 Chinese had been wounded. Although the South Koreans frequently called down artillery support during the attack, most of the casualties inflicted on the enemy were from rifle and machine gun fire. The 17 April probe was to mark the last major infantry action for the 1st Marine Division during its second month on JAMESTOWN.

Throughout the month a total of 5,000 rounds of artillery fire and 3,786 rounds of mortar fire fell in the division sector. On 2 April the greatest volume for any single day was received: 3,000 artillery and 118 mortar rounds. An average day's incoming, during April, was approximately 167 artillery and 125 mortar rounds.

Strengthening the Line[49]

Even before the Communists had launched their mid-April attacks against JAMESTOWN, the 1st Marine Division had implemented plans

---

[47] *Culhane ltr.*

[48] LtGen Merrill B. Twining ltr to Deputy Asst CofS, G-3, HQMC, dtd 19 Aug 54.

[49] Unless otherwise noted, the material in this section is derived from: *PacFlt Eval Rpt* No. 4, Chap. 9; 1stMarDiv ComdD, Apr 52; KMC Regt UnitRpt 46, dtd 17 Apr 52.

to strengthen its line in western Korea. Besides the digging, timbering, and sandbagging to accomplish a major improvement of the physical defenses, General Selden required Marine infantry regiments to conduct an aggressive defense of their sector of responsibility. He ordered MLR units to employ snipers all along JAMESTOWN and to dispatch daily patrols forward of the line to ambush, raid, kill, or capture Chinese and their positions. The division commander further directed that supporting arms such as artillery, tank, and air, when available, be used to destroy hostile defenses, harass the enemy, and break up his assemblies as well as to protect Marine positions.

As a result of an I Corps directive, the 1st Marine Division assumed responsibility for an additional 6,800 yards of front on 14 April from the 1st Commonwealth Division sector to the right of the division. In preparation, the 5th Marines had taken over the western end of the 1st Marines sector, held by 2/1, two days earlier. On the 14th the 1st Marines, newly commanded by Colonel Walter N. Flournoy,[50] extended its line eastward to assume new limiting points and part of the MLR in the western part of the Canadian Brigade sector. Relief of the Commonwealth unit was completed without any difficulty or enemy interference. This additional yardage, plus the Kimpo Peninsula front, now stretched the Marine division MLR to 35½ miles.

As a result, General Selden found it necessary to withdraw the division general outpost line in order to build up his main line of resistance. On 17 April, the 1st KMC Regiment reduced its OPLR to an OPLO (outpost line of observation) and the left battalion pulled its MLR back to more defensible ground. The Marine division center and right regiments withdrew their outpost lines on 23 and 24 April. Both regiments then established forward outposts and listening posts which, in many cases, utilized former OPLR positions. Many of these posts were manned during daylight hours only.

Abandonment of the forward OPLR added strength to the main line, but it also meant that frontline battalions had to commit all their companies on line, thus losing their reserve. To prevent Chinese occupation of desirable terrain features on the former OPLR, the division dispatched combat and reconnaissance patrols forward of its

[50] Colonel Flournoy became regimental CO on 10 April, succeeding Colonel Wade.

line. In the KMC sector, the only Marine area favorable for tank operations forward of JAMESTOWN, tank-infantry patrols were periodically employed.

To the west of the KMC sector, the Marine 1st Amphibian Tractor Battalion (Lieutenant Colonel Michiel Dobervich) was assigned a section of the KANSAS Line to defend, beginning 16 April. Reinforced by attachment of the Division Reconnaissance Company (Major Ephraim Kirby-Smith) that same day, Lieutenant Colonel Dobervich employed Company C (two platoons), the headquarters LVT platoon, and the reconnaissance unit to man 30 defensive positions from the Han River eastward to the KMC western boundary.[51]

Two other measures to strengthen his sector of JAMESTOWN were utilized by the Marine division commander. On 18 April, he asked General O'Daniel to reconsider the no-fire zone recently established by the corps commander. General Selden, who had received reports of Chinese use of the sanctuary located within Marine Corps territory —for firing positions and assembly areas primarily—recommended, after I Corps had refused him permission to fire into the haven, a redrawing of the O'Daniel line to coincide more closely with the boundaries established by the UN. Approval along the lines submitted by the division was given by I Corps that same day. The second measure employed by General Selden was use of an additional defensive line, WYOMING FORWARD. This position, closely paralleling JAMESTOWN in the KMC and 5th Marines sectors, added depth to the sector defenses.

A unique rescue and recovery operation also came into existence about this time. On 19 April the division ordered the 5th Marines, occupying the center regimental sector, to organize a tank-infantry force for rescue of the United Nations Truce Team, should such action become necessary. The regimental plan, published on 22 April, utilized a reinforced rifle company-tank company organization directly supported by organic 5th Marines 4.2-inch mortars and 1/11. The Everready Rescue Force, from the regimental reserve, occupied the high ground (OP 2) east of and dominating Panmunjom.

In addition to setting forth organizational details of the task unit, the 5th Marines Operational Plan 6–52 specified the method of operation for the rescue force. Taking advantage of the peace corri-

---

[51] Company A, 1st Amphibian Tractor Battalion had been attached to the Kimpo Provisional Regiment since 31 March and Company B was supporting MAG–33 at Pohang.

dor in the western end of the center sector, a Forward Covering Force would speed tank-riding infantry to the high ground one-half mile beyond the objective, Panmunjon. Following would be the Pick-Up Force, from the 1st Tank Battalion Headquarters Platoon, which would retrieve the principal UN delegates and take them quickly to the assembly area two miles to the rear of the MLR. A Rear Covering Force, composed of a tank-infantry element, would follow the Pick-Up force both on its way towards the objective and on the return trip. Withdrawal of both covering forces was regulated by a series of phase lines.

Marine Air Operations[52]

Even though the Marine air-ground team had been shorn of much of its tactical aviation, what remained was well utilized. Helicopter troop operations had become commonplace by the end of April 1952. That month there were three exercises to further evaluate tactical concepts of helicopter employment. Operation PRONTO, conducted on 5 April, was the first major troop lift in the new I Corps sector. In this maneuver approximately 670 troops of 2/7 and 10,000 pounds of rations were transported by helicopter and truck from the Munsan-ni vicinity across the Han River to the Kimpo Peninsula. Here the reserve battalion served as a counterattack force in a hypothetical enemy landing. Due to the necessity for avoiding the neutrality zone in the Munsan area, round-trip flights averaged about 57 miles.

The exercise combined the shortest notice and longest distance of any large-scale helicopter troop movement conducted by HMR–161. It pointed to the fact that a helicopter unit could successfully lift a troop organization virtually as an "on call" tactical tool and without the benefit of previous liaison.

Operation LEAPFROG, on 18–19 April, transported one KMC battalion across the Han to the peninsula and lifted out another the following day. The purpose of this test was to determine the feasibility

---

[52] Unless otherwise noted, the material in this section is derived from: *PacFlt Eval Rpt* No. 4, Chaps. 9, 10; 1st MAW, HMR–161, VMO–6 ComdDs, Apr 52; Lynn Montross, *Cavalry of the Sky—The Story of U. S. Marine Combat Helicopters* (New York: Harper & Brothers, 1954), hereafter Montross, *SkyCav*, quoted with permission of the publishers.

of a replacement movement conducted over water, with "consideration given to the language barrier existing between the troops and the transporting facility." [53] The six-mile round trip was the shortest troop haul yet made by the transport chopper squadron. Consequently, it took the 12 HRS–1 single-engine Sikorsky aircraft only 3 hours and 26 minutes to complete the exchange of the 1,702 KMC troops.

Colonel McCutcheon's HMR–161 pilots found that their helicopters could carry six combat-equipped Korean Marines instead of five American Marines, due to the smaller size and weight of the average Korean. Since the U.S. and KMC Marine battalions were the same size, the larger load factor for the Korean Marines enabled their unit to be moved faster. In LEAPFROG the language difference proved to be no handicap, since there were sufficient interpreters on hand and the troops were cooperative. Helicopter pilots could use landing sites close together because the terrain was open and the area of operations beyond the reach of Chinese artillery.

Close on the heels of LEAPFROG came a third airlift. Operation CIRCUS, conducted on 23 April, provided for the air deployment of the 7th Marines reserve regiment, minus two battalions, across the Imjin to landing sites just to the rear of the secondary defensive line, WYOMING FORWARD. Ten helicopters carried 1,185 Marines over the river barrier to blocking positions in 90 minutes. The CIRCUS exercise illustrated that a minimum distance should be maintained between loading and unloading sites for a safe and efficient transport operation. It also pointed up that "consideration must be given to the number of aircraft assigned to each traffic pattern during short hops over a river." [54] This successful maneuver came three days before all HRS–1 aircraft were grounded due to a defect in the tail rotors. By mid-May the problem had been corrected and the aircraft returned to flying status.

During April, Lieutenant Colonel Herring's VMO–6 employed its 11 single-engine OE–1 observation planes for a total of 508 fixed-wing combat flights. More than half of these, 275, were for artillery spotting; of the remainder, 166 were flown for reconnaissance and 67 represented photo, weather, liaison, and area check-out maneuvers.

---

[53] *PacFlt EvalRpt* No. 4, 10–73.
[54] *PacFlt EvalRpt* No. 4, p. 9–50.

Combat flights by the squadron helicopters[55] during the month were 110 liaison, 45 reconnaissance, and 93 evacuations. Of the total 756 combat flights performed by both fixed-wing and rotary craft, 511 were over enemy territory.

During that same month, Marine squadrons operating under the Fifth Air Force put a total of 2,708 planes into the air despite restrictive or prohibitive weather on 20 days. Continuing its emphasis on attacking the North Korean transportation system, the Air Force command dispatched 1,397 Marine planes on interdiction missions. Marine-piloted close air support sorties flown to assist the 1st Marine Division numbered only 56 throughout April; those piloted by Marines for 16 other UN divisions totaled 547.

Not all the air sortie records were made by land-based Marine squadrons. On 18 April, VMA–312, the CTE 95.11 squadron provided by the 1st Marine Aircraft Wing, flew 80 sorties, a Korean record for a carrier-based squadron to that date and twice the daily average for the initial six months of 1952.

By 20 April the three tactical squadrons of MAG–12—VMF(N)–513, VMF–212, and VMF–323—had completed their relocations on the Korean west coast. Two days later, combined MAG–12 attack and –33 jet aircraft participated in what was a Fifth Air Force one-day combat record: 1,049 sorties.

One MAG–33 unit, the newly-formed Marine Photographic Squadron 1, was already flying a large number of aerial reconnaissance missions directed by Fifth Air Force. It provided almost one-third of the daylight photo effort required by FAF with but one-quarter of the aircraft.[56] VMJ-I's complement of a dozen 550 mph McDonnell twin-jet Banshee F2H-2P aircraft mounted three cameras and were capable both of high altitude work and good speed. Introduction of this single-seat jet was considered the "first important develop-

---

[55] Rotary wing aircraft assigned were two types, HTL–4 and HO3S–1. The former is a two-place, plastic-dome Bell product; the latter, the first helicopter operated by the Marine Corps, is an observation-utility, three-passenger Sikorsky-made craft. HistBr, G–3 Div, HQMC, *Marine Corps Aircraft, 1913–1965,* Marine Corps Historical Reference Pamphlet (Washington: 1967 ed.) pp. 34, 38.

[56] *PacFlt EvalRpt* No. 4, pp. 10–2, 10–108. This record was established despite the fact that the Marine squadron, with 10 jets, flying out of K–3 (Pohang) was more than 150 miles further from most targets than the other major photo unit, the 15th Tactical Reconnaissance Squadron of the 67th Tactical Reconnaissance Wing, based at K–14 (Kimpo).

ment in aerial photography in the Korean War."[57] since the Banshee could outproduce any photo plane in Korea.

The month of April also marked change of command ceremonies for the 1st Marine Aircraft Wing. On 11 April at K–3, General Schilt turned over wing responsibility to Brigadier General Clayton C. Jerome. Among the numerous civilian and military dignitaries attending the ceremony at the Pohang 1st MAW headquarters were the Honorable John J. Muccio, U.S. Ambassador to Korea; Air Force Lieutenant Generals Otto P. Weyland and Frank F. Everest, commanders of FEAF and FAF respectively; and the Marine division CG, Major General Selden.

The new wing commander, General Jerome, like his predecessor, had a distinguished flight career. A 1922 graduate of the Naval Academy, he had served in various foreign and U.S. aviation billets and was a veteran of five World War II campaigns. In 1943 Colonel Jerome was operations officer for Commander, Aircraft, Solomon Islands. Later he was named Chief of Staff, Commander, Aircraft, Northern Solomons and Commander, Aircraft and Island Commander, Emirau, in the northern Solomons. Before returning to the States, Colonel Jerome had participated in the recapture of the Philippines, commanding MAG–32 and directing all Marine air support in the Luzon fighting. Brigadier General Jerome became Director of Aviation and Assistant Commandant of the Marine Corps for Air in September 1950 and served in this capacity until taking command of the 1st Marine Aircraft Wing in Korea.[58]

During the command ceremonies the outgoing 1st MAW commander, General Schilt, was presented the Distinguished Service Medal for his outstanding leadership of the wing. The award was made by Lieutenant General Weyland. Shortly before his Korean tour ended, General Schilt had also received from ROK President Syngman Rhee the Order of Military Merit Taiguk, for his contribution to the military defense of South Korea.

---

[57] *Ibid.*, p. 10–59.

[58] DivInfo, HQMC, Biography of LtGen Clayton C. Jerome, Jul 58, rev.

Supporting the Division and the Wing[59]

Because of the command relationships existing in Korea, with all ground units under operational control of CG, EUSAK, the majority of the logistical support to the Marines was handled by the Army. Eighth Army, 2d Logistical Command (2d LogCom) provided for resupply of items used commonly by both Marine and Army personnel; the Marine Corps (Commanding General, FMFPac) furnished those supplies and equipment used by Marine units only.

When the division moved to the west, the 1st Shore Party Battalion opened a rear service area at Ascom City. Here the division established and maintained Class II (organizational equipment) and IV (special equipment) dumps for its units, as well as Class I (rations) and III (petroleum products) facilities for both the Kimpo regiment and the service units stationed at Ascom. Class I shipments were forwarded to the Munsan-ni railhead and stored there. Fuels and lubricants and Class V items (ordnance) were received from the U.S. Army. A forward ammunition supply point (ASP) was located north of the Imjin to assure a steady flow of ammunition to frontline combat units in the event that either an enemy attack or emergency flooding conditions of the river prevented use of the bridges. For the same reason a truck company was positioned near this supply point each night.

Reinforcing the division logistic effort was the 1st Combat Service Group. Commanded by Colonel Russell N. Jordahl, the 1st CSG in late April had nearly 1,400 Marines and Navy medical personnel stationed at various points between Japan and Korea. At Kobe, Japan, the Support Company processed Marine drafts arriving and departing Korea. At Masan, the Supply Company, 1st CSG, requisitioned for the division those Class II and IV items peculiar to the Marine Corps needs and forwarded them upon request. Heavy maintenance of all technical equipment was performed by the Maintenance Company. Supporting the 1st Motor Transport Battalion operation was the Motor Transport Company, 1st CSG. Most of the group, including Headquarters Company, was based at Masan.[60] Splinter detach-

---

[59] Unless otherwise noted, the material in this section is derived from: *PacFlt EvalRpt* No. 4, Chaps. 9, 10; 1stMarDiv, 1st MAW, 1st CSG, 11thMar, 1st TkBn ComdDs, Mar–Apr 52; 1st CSG UnitRpts, Apr 52.

[60] The Support Company moved to Ascom City on 14 Jun 52.

ments from the group also operated transport facilities at other locations in Korea.

In western Korea, good rail transport into Munsan-ni and an adequate but not all-weather road system improved the division's logistical situation. Greater storage facilities also existed in the JAMESTOWN rear supply areas than in the X Corps sector just vacated by the Marines. Division motor equipment did not suffer any appreciable damage due to the rigors of the MIXMASTER transplacement. Vehicle maintenance also presented a favorable outlook, due to the expected decreased use during the period of positional warfare. On the other hand, an unduly large number of tanks developed engine troubles in March, which were traced back to defective oil cooling fans. This condition was corrected in April and May by installation of new fan assemblies.

Guns of the 1st Tank Battalion immediately began to render valuable support to Marine frontline regiments with the division's new assignment in the west. Companies A, B, and C were placed in direct support of the three forward infantry regiments. Company D drew the reserve mission, which included tank-infantry training with the 7th Marines and preparation for reinforcing division artillery fires. Tank companies were used almost daily in the forward sectors for destruction by direct fire of the Chinese MLR fortifications. For such missions the M–46 tanks, equipped with high-velocity 90mm guns, lumbered forward from secure assembly areas to the rear of JAMESTOWN to temporary firing positions on the line.

After pouring direct fire on preselected targets and completion of the fire mission, the armored vehicles then returned to the rear. Less frequently, a five-vehicle tank platoon accompanied a reinforced rifle platoon and conducted daylight reconnaissance missions of forward areas to engage the Chinese and to gain intelligence about enemy positions and terrain. During April six such tank-infantry patrols, all in the KMC regimental area, failed to establish direct contact with the enemy but did draw mortar and artillery fire.

Marine artillery, which had been receiving its share of attention from Communist field guns,[61] was faced by problems in two other

[61] One artillery weapon, in particular, as well as the Marine tanks habitually drew the fury of Chinese counter-fire. The heavy destructive power of the U.S. Army 8-inch, self-propelled howitzers firing on tough Chinese defensive positions, generally brought down on their own emplacements a rain of enemy shells, so sensitive were enemy commanders to these hard-hitting weapons. *Pala comments.*

respects. Although the enemy held only four more artillery weapons than did the Marines, General Selden still lacked the ability to mass artillery fires to the same degree as did the Chinese.[62] This limitation stemmed directly from the wide physical separation of 11th Marines batteries and the frontline infantry regiments being supported. A second problem, the loss of qualified forward observers—reserve officers due to return to the States for release from active service—forced the 11th Marines to begin a school to train infantry officers for this function. To make the course realistic, all firing was done at live targets.[63]

In April 1952, the 11th Marines organization had three light 105mm howitzer battalions (54 guns), one medium 155mm howitzer battalion (12 guns), the KMC 105mm howitzer battalion (18 pieces), and a 4.5-inch rocket battery (6 launchers). Attached to the 1st Marine Division and located in its sector were one battalion and one battery of the I Corps field artillery. The mission of the Marine artillery regiment was to provide accurate and timely fires in support of both the MLR and OPLR defenses, until withdrawal of the latter late in April. Batteries of the 11th Marines also fired on known and suspected Chinese gun emplacements and on targets of opportunity. The regiment also provided intelligence on enemy artillery.

Throughout April, Colonel Henderson's units continued to improve their tactical and administrative areas, concentrating on field fortifications, wire communications, and road trafficability. In the last category, the artillery dozers and dump trucks not only did nearly all of this work for the 11th Marines but also provided "a fair amount of 'direct support' bulldozing to the infantry regiments and occasionally loaned dozers and operators to the engineers." [64]

Within a Marine aircraft wing, personnel and equipment for logistic support are purposely limited to carrying out the wing primary mission—providing air support during an amphibious operation. The wing T/O (Table of Organization) provides a streamlined organization with light, transportable organic equipment. Additional logistical support personnel and equipment are not included since this would result in (1) a duplication of support effort between the wing

---

[62] *PacFlt EvalRpt* No. 4, Chap. 9, p. 9–39.
[63] BGen Frederick P. Henderson ltr to CMC, dtd 6 Sep 66, hereafter *Henderson ltr II.*
[64] *Ibid.*

and landing force and (2) a great increase in wing transport shipping requirements. When the wing moves ashore, organic units render support necessary for operations on the airfield only. Responsibility for activities beyond this basic mission—airfield construction, maintenance of runways, and movement of supplies to the airfield—must come from more senior commands. Usually such assistance is obtained by attaching elements of a naval construction battalion and other logistical support units.

In April 1952, Naval Construction Battalion Unit 1804 assisted in the construction and maintenance received by MAG–33 at K–3. Here at the port of Pohang, a detachment from the 1st Combat Service Group controlled the movement of fuels, oils, lubricants, and ordnance to wing dumps. Amphibian tractors (LVTs) of Company B, 1st Amphibian Tractor Battalion, provided most of the transportation required for these supplies excepting ordnance. Assistance in the form of amphibious trucks (DUKWs) was furnished by a platoon from the 1st Amphibian Truck Company. When required, Marines of these two companies manhandled the supplies.

Logistical support for the Marine wing was governed by the same general procedures that applied to the division; 1st MAW supply requirements beyond its augmented capability became the responsibility of Eighth Army (2d LogCom) which furnished items common to both Marine and Army units. If this EUSAK agency did not stock the requisitioned item, it provided a substitute. Responsibility for resupply of aviation items rested with the U.S. Navy. Commander, Naval Forces, Far East (ComNavFE) replaced unserviceable aviation technical equipment such as aircraft parts and special maintenance tools. Commander, Service Force, Pacific (ComServPac) replenished aviation ordnance. Responsibility for supplying items peculiar to the Marine Corps rested with CG, FMFPac.

The repair and maintenance of 1st MAW equipment posed far less of a problem than the construction and upkeep of airfields. Major repair work on aircraft was satisfactorily performed in Japan by the wing support squadron at Itami, and by the U.S. Navy Fleet Air Service Squadron 11 (FASRon–11), located at the Naval Air Station, Atsugi. The establishment in Japan of the wing heavy maintenance facility depended, in part, upon its proximity to the wing flying squadrons. Other considerations were the availability to the wing commander of adequate air transport for continuous resupply of both routine and emergency items and reliable communications between the users and the maintenance unit. Because these conditions favoring removal of the heavy maintenance facility from the immediate combat area existed throughout Korean hostilities, it was possible for the maintenance units to operate successfully in Japan away from the combat zone.

Air base construction and maintenance of airfield runways and taxiways had plagued wing operations since the early days of the Korean War. During the first winter these problems had appeared repeatedly at those installations where Marine air was either not properly supported or insufficiently augmented by the operational commander. Shortly after MAG–33 had moved to K–3 in early 1951, the wing commander requested emergency repairs for the runway and a permanent solution to the airfield maintenance difficulties. Assistance was made available, but it was insufficient. The repair force had to be augmented by Marines pulled away from their own vital jobs and by native laborers. Later, in the spring of 1952, when the Air Force assigned some of its engineers to assist, the maintenance problem almost disappeared.

Motor transport within the wing was a continuing source of logistical problems. Vehicles for handling the heavier aviation ordnance were unsatisfactory because their configuration, of World War II vintage, did not permit them to service the newer aircraft. Other trucks lacked engine power or rigidity to withstand sustained use under primitive airfield conditions. World War II vehicles that had been preserved and placed in open storage required reconditioning before their use in Korea. Mechanics' general and special tools had a high replacement rate throughout the entire period of wing operations in Korea.

Aircraft fuel handling in April 1952 followed outmoded World War II methods. For K–3, amphibian vehicles received drummed fuel from ships and landed it at the beach. There MAG–33 personnel transferred the gasoline to 1,200-gallon fuel trucks, which then moved it to the airfield servicing area, where other Marines transferred it again, this time to 3,200-gallon stationary refuelers for dispensing into the aircraft. Although this method became highly developed, it was extremely slow and wasteful of manpower and vehicles in comparison to the tank farm system, which was soon to reach K–3.

Two areas of logistics continued to remain almost trouble free for division and wing Marines. Medical problems existed but were not extensive. During a five-day period in late March, Marine Air Control Group 2 experienced 13 cases of scarlet fever but no fatalities. That same month, the Pacific Fleet Medical Officer noted that MAG–12 sick bays were in excellent condition and that medical "personnel have shown great ingenuity in fabricating various items of medical equipment from scrap metal and lumber."[65]

Evacuation of casualties and the utilization of air vehicles for transport of passengers and cargo proved to be the second asset in logistical operations. The Itami-based detachment of VMR–152 moved 7,757 personnel from the division and wing and 738.7 tons of cargo during April 1952. In addition, the R5D craft hauled a total of 325.2 tons of U.S. mail that month for the two Marine organizations. Speedy removal of patients to better equipped facilities in the rear by VMO–6 and HMR–161 helicopters was a giant step forward in life-saving techniques. VMO–6 usually provided this service, but early in April, Colonel McCutcheon's squadron was assigned emergency medical evacuation duties to augment the observation squadron.[66] Pilots flew these evacuation missions with almost total disregard for adverse weather or darkness, and without radar control or adequate instrumentation for all-weather operations.[67]

Different Area, Different Problem[68]

An additional responsibility the 1st Marine Division inherited when it moved to western Korea was control of civilians wthin the division boundary. In eastern Korea, all nonmilitary personnel had been evacuated from the vicinity of the MINNESOTA Line in the division sector; they had not been removed from the JAMESTOWN area. Prior to the

[65] *PacFlt EvalRpt* No. 4, Chap. 12, p. 12–8. The medical officer's report to CinCPac noted that a vast improvement "in the spaces allocated for the care of the sick and wounded" had been made.
[66] *PacFlt EvalRpt* No. 4, p. 10–69, p. 10–73.
[67] *Ibid.*, p. 10–68. Flights were not made in heavy fog. Test use by the Marine Corps Equipment Board of some of the equipment needed to navigate under conditions of reduced visibility was nearing the end of its development cycle.
[68] Unless otherwise noted, the material in this section is derived from: CG, 1stMarDiv ltr to CMC, dtd 23 Jul 53, Subj: Type "C" Rpt: "Civilian Affairs and the Korean Service Corps, Mar 52–May 53," hereafter CG, 1stMarDiv ltr, *Civ Afrs and KSC;* 1stMarDiv ComdDs, Mar–Apr 52; HqBn, 1stMarDiv ComdDs, Mar–Apr 52.

arrival of the division in the west, the STAYBACK Line, averaging seven miles to the rear of the Imjin River and running in a generally northeast-southwest direction, had been established to limit the movement of civilian personnel in the forward areas. The Marines soon found that their predecessors must have been lax, however, in requiring that Korean civilians remain behind STAYBACK. What seemed equally unsuitable to the division was the poor military-civilian relationship that had apparently existed for some time.

To correct the situation, General Selden cautioned his units to avoid unnecessary damage or destruction to the civilian economy. He directed his commanding officers to keep unauthorized Koreans away from Marine installations. Military police set up check points and instituted roving patrols to enforce division controls. Civil violators were turned over to Korean authorities or held for investigation before release. Civilians who lived in the forward areas were removed to the rear. They were prevented from going beyond STAYBACK until August 1952, when a controlled passage system was instituted.

CHAPTER II

Defending the Line

UN Command Activities—Defense of West and East Coast Korean Islands—Marine Air Operations—Spring 1952 on JAMESTOWN*—End of the Second Year of War—A Long Fourth of July—Changes in the Lineup—Replacement and Rotation—Logistical Operations, Summer 1952*

UN Command Activities[1]

MOVEMENT OF the 1st Marine Division to the west was part of an Eighth Army master plan to strengthen UN defenses and at the same time to enable South Korean forces to assume increased responsibility in the defense of their homeland. The tactical realignment in the spring of 1952 put more South Korean infantry units on the main line of resistance and buttressed the fighting front with five corps sectors instead of four. In the far west, the I Corps positions were newly manned (left to right) by the 1st Marine, 1st Commonwealth, 1st ROK, and the U.S. 45th Infantry Divisions. Next in line was IX Corps, whose left boundary General Van Fleet[2] had shifted further west, which now had a divisional line up of the ROK 9th on the left, the U.S. 7th in the center, and the U.S. 40th on the right.

[1] Unless otherwise noted, the material in this section is derived from: Cdr Malcolm W. Cagle, USN and Cdr Frank A. Manson, USN, *The Sea War in Korea* (Annapolis, Md.: U.S. Naval Institute, 1957), hereafter Cagle and Manson, *Sea War, Korea;* James A. Field, Jr., *History of United States Naval Operations, Korea* (Washington: [Div. of Naval Hist], 1962), hereafter Field, *NavOps, Korea;* John Miller, Jr., Maj Owen J. Carroll, USA, and Margaret E. Tackley, *Korea, 1951–1953* (Washington: OCMH, DA, 1958), hereafter Miller, Carroll, and Tackley, *Korea, 1951–1953.*

[2] General Van Fleet, CG, EUSAK since April 1951, had advocated a program in which South Korean troops would be rigorously trained to take over an increasingly greater part of the UNC defense efforts in Korea. See Mark W. Clark, *From the Danube to the Yalu* (New York: Harper & Brothers, 1954), p. 185, hereafter Clark, *Danube to Yalu,* quoted with permission of the publishers.

To fill in the central part of the EUSAK front where the change of IX Corps boundary had created a gap in the line, the UN commander inserted the ROK II Corps with three divisions (ROK 6th, ROK Capital, and ROK 3d) forward. Immediately to the right of this new ROK corps sector, the X Corps continued in approximately its same position on the east-central front. Its ROK 7th and U.S. 25th Divisions remained on line, while the ROK 8th had advanced to the former sector of the Marine division in the wild Punchbowl country. At the far right of the UN line, the ROK I Corps front was held by the ROK 11th Division at the X Corps boundary and the ROK 5th along the Sea of Japan. By 1 May 1952, nine Republic of Korea divisions had been emplaced on the UNC main defense line, three more than had been there in mid-March.

Throughout Korea in March and April there had been a general stagnation of offensive action on both sides because of fog, rain, and mud. In May, however, the Chinese launched no less than 30 probing attacks against the ROK 1st Division in the I Corps sector, without gaining any significant advantage. To the right, the enemy and the U.S. 45th Division traded blows in several patrol actions. In June, major EUSAK combat action was still centered in the 45th's sector, but the following month was marked by sharp battlefront clashes in nearly all Eighth Army division areas. For a two-week period in July and August, heavy seasonal rains limited both ground and air action. With the return of normal weather, heavy fighting again broke out, this time concentrated in the I Corps sector. This action did not abate until late August, when the onset of the heaviest rains of the season again drastically reduced military operations.

Communist ground activity in the spring of 1952 was marked by increased artillery support which resulted in telling damage to UN infantry and artillery positions. Thus, during May, the enemy expended approximately 102,000 artillery and mortar rounds against the Allied front, roughly 12 times the number fired the previous July, just prior to the period of stabilized battlelines in Korea. The artillery buildup was accompanied by a sharp decrease in hostile air support activities. While the Chinese had flown 3,700 jet sorties during the first month of 1952, by June the monthly total had dropped to 308.

As part of the balanced military forces, Allied air and sea units continued their active defense in support of UN ground units. Beginning in late May, Fifth Air Force shifted the emphasis of its destructive effort from interdiction of communication routes to the bombing of selected industrial targets. Naval air was committed to support the FAF programs. At sea, ships steamed almost at will to sustain the U.S. lifeline. Underscoring the complete UN control of Korean waters, large naval vessels offshore fired their big guns in support of ground troops. Off both the west and east coasts, Task Force (TF) 95 maintained its blockade of North Korean ports and reduced the extent of water travel that enemy craft could safely undertake. This same naval force was responsible for the Allied defense of islands located off the east and west coasts of Korea.

Defense of West and East Coast Korean Islands[3]

Just off the northwest Korean mainland a string of islands extends from the mouth of the Yalu River down around the peninsula to Pusan in the southeast. Most of these islands are tiny and are located south of the 38th Parallel. Only a few lie off the east coast, and these are clustered primarily in the North Korean harbor of Wonsan. By early 1951, UN forces exercised control over most of the Korean islands. Their tactical importance is shown from their diverse use as sites for UN Command intelligence activities, USAF radar installations, locations for the emergency landing strips used by Allied planes, bases for U.S. search and rescue operations, and as springboards for possible thrusts into enemy rear areas.[4]

Another reason for holding some of the islands had come to light during truce negotiations in December 1951. At that time, in an attempt to expedite the successful conclusion of the truce meetings, UN representatives had offered the Communists all the islands north of the 38th Parallel. Brushing aside the tactical value of the proposal, the enemy boasted that he could capture the islands at any

[3] Unless otherwise noted, the material in this section is derived from: *PacFlt EvalRpts* No. 4, Chap. 9; No. 5, Chap. 8; West Coast Island Defense Element ComdDs, Feb–Oct 52, hereafter *WCIDE ComdD,* with date; East Coast Island Defense ComdDs, Jan–Oct 52, hereafter *ECIDE ComdD,* with date; Col William K. Davenport ltr to CMC, dtd 27 Jun 52, Subj: Type D Report of duty as Commander West Coast Island Defense Element (CTE 95.15); Cagle and Manson, *Sea War, Korea;* Field, *NavOps, Korea.*

[4] Evidence of Chinese concern about such rear area attacks is apparent in the countermeasures taken: "Order of Battle reports indicated that a total of three North Korean Corps and three Chinese Communist Armies were engaged in coastal defense operations on the east and west coasts of North Korea." *PacFlt EvalRpt,* No. 5, p. 8–79.

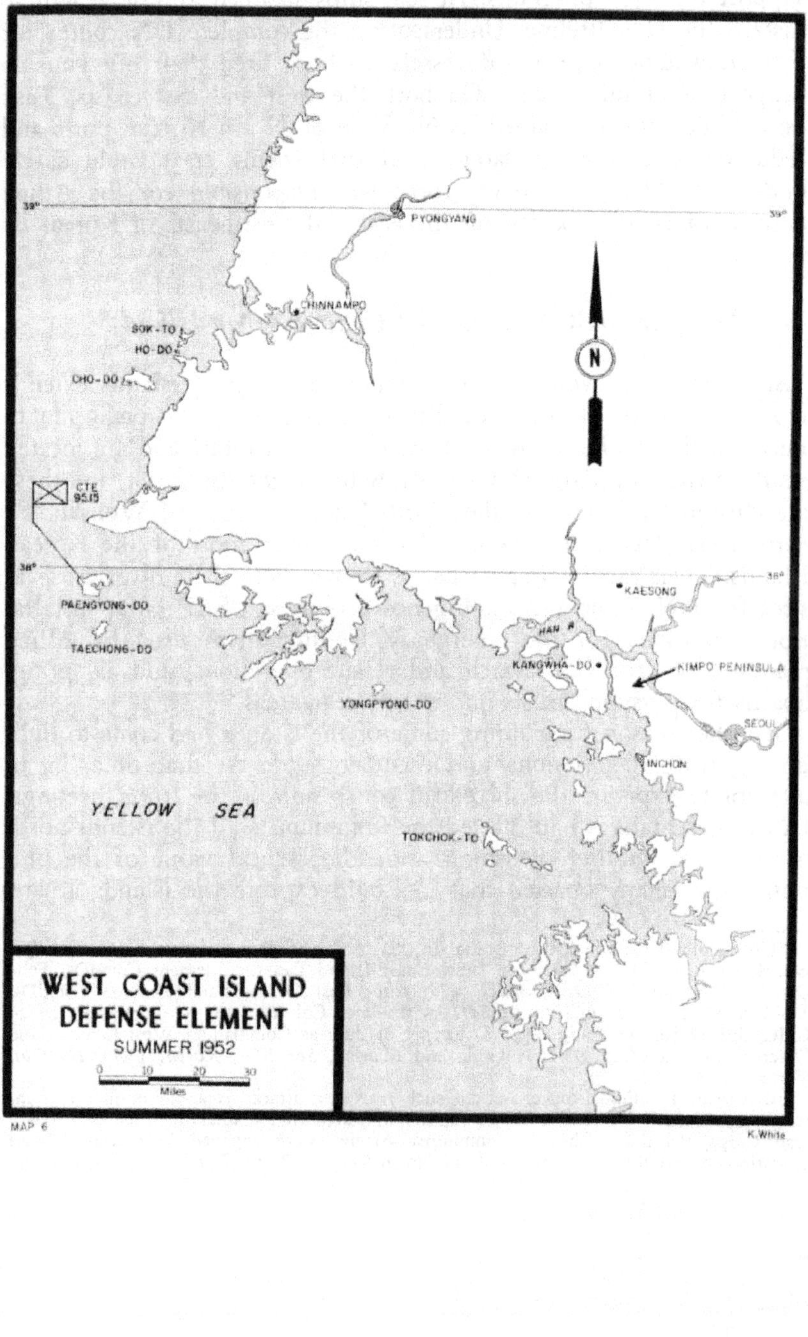
39°
PYONGYANG
CHINNAMPO
N
SOK-TO
HO-DO
CHO-DO
CTE 95.15
38°
PAENGYONG-DO
TAECHONG-DO
KAESONG
HAN R
KANGWHA-DO
KIMPO PENINSULA
YONGPYONG-DO
SEOUL
INCHON
YELLOW SEA
TOKCHOK-TO
WEST COAST ISLAND DEFENSE ELEMENT
SUMMER 1952
0 10 20 30
Miles

MAP 6

K.White

time. In November 1951 the Communists had, in fact, seized two western islands near the mouth of the Yalu. The 1,000 defending guerrillas there—former North Koreans working for the UNC—had been unable to stem the assault. The UN Command promptly reviewed the island situation and on 6 January 1952 gave TF 95, the United Nations Blockading and Escort Force, responsibility for both overall defense and local ground defense for the 11 coastal islands north of the 38th Parallel and the 4 islands immediately south of this boundary. Two subordinate blockade task groups, one in the west and another in the east, were responsible for the defense of these islands.

In the west, Task Group (TG) 95.1 was charged with the defense of six islands. (See Map 6.) Two of these, Sok-to and Cho-do, lie between the 38th and 39th Parallels; the four remaining islands, Paengyong-do, Taechong-do, Yongpyong-do, and Tokchok-to, are above the 37th Parallel. In the east, TG 95.2 was responsible for keeping nine islands north of the 38th Parallel in friendly hands. Situated in Wonsan harbor are Mo-do, Sin-do, So-do, Tae-do, Hwangto-do, Ung-do, and Yo-do, the largest. (See Map 7.) Another island, Yang-do, actually a two-island group further north in the area of the 41st Parallel, is 18 miles northeast of the coastal city of Songjin. The southernmost island, tiny Nan-do, is below Wonsan and the 39th Parallel and lies 10 miles northeast of Kojo, another coastal city.

Ground defense of the islands had been, at best, a haphazard arrangement before TF 95 took over the responsibility. Many of the islands, especially those inhabited by friendly guerrillas, had neither plans for a proper defense nor commanders experienced in organizing resistance to enemy attack. Soon after the two islands near the mouth of the Yalu were taken, ROK Marines were rushed to those islands considered most strategic for South Korean defense. Late in 1951, U.S. Marines had been assigned to the area in an advisory capacity. By early 1952, Marine Corps detachments were in command of the island defense activities for both task groups. Korean Marines provided a majority of the actual defending forces.

Although the 1st Marine Division initially had supplied the officers and men for the island security missions, in January 1952 FMFPac took on direct responsibility for furnishing personnel and providing for their administrative and logistical support through the 1st Provisional Casual Company, FMFPac. Located at Otsu, Japan, the company was the administrative headquarters for seriously wounded Marine division and wing personnel recuperating in service hospitals in Japan. Recovered patients who volunteered for duty with the offshore commands provided the bulk of the Marines used in this defense. Major responsibilities were to plan, organize, and conduct the defense of these islands off the Korean west and east coasts. A task element under each task group was created for this purpose.

With its headquarters at Paengyang-do, Task Element (TE) 95.15, the West Coast Island Defense Element (WCIDE), was organized early in January 1952. The following month, the initial complement of U.S. Marines arrived. Colonel William K. Davenport, Jr., element commander, assigned his 5 officers and 29 enlisted men to the 4 most critical islands and to his staff. Those islands garrisoned were Cho-do and Sok-to, north of the Parallel and both within range of enemy mainland guns, and Paengyang-do and Yongpyong-do, to the south. Taechong-do, near the command island, and Tokchok-to, southwest of Inchon, were both considered secure and not provided with U.S. Marine commanders. At each of the four occupied islands, Marines reconnoitered the terrain, drew up plans for preparation of defensive positions, organized and trained the troops available, and began the laborious task of constructing the defense. Protection against long-range hostile artillery fire was emphasized for the northern Sok-to and Cho-do garrisons.

Off the other long coast of Korea, TE 95.23, the East Coast Island Defense Element (ECIDE), commanded until early May 1952 by Colonel Frank M. Reinecke, had an almost entirely different situation. Eight of the nine islands in the vicinity of Wonsan Harbor or north of Songjin that ECIDE was responsible for were within range of Communist shore batteries and thus frequently fired upon. Even before the January 1952 decision, the U.S. Navy had been charged with the security of these east coast islands north of the 38th Parallel. For these reasons ECIDE defenses had to maintain a greater state of readiness and were more advanced than in the west. Fire support ships and land based U.S. Marine naval gunfire spotting teams from 1st ANGLICO (Air and Naval Gunfire Liaison Company), FMF, which also provided forward air controllers for the KMC regiment, stood by at all times to silence unfriendly artillery fire emanating from the mainland. The Marines had also trained Korean Marines to handle the spotting missions.

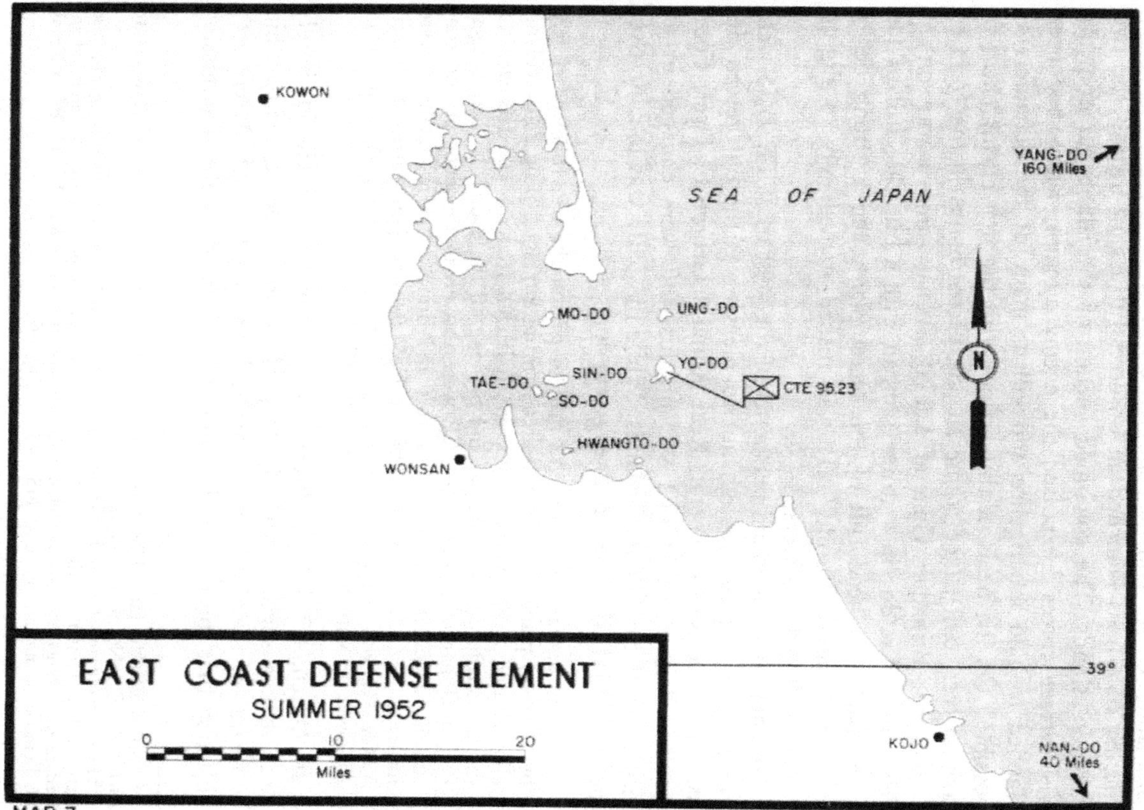
EAST COAST DEFENSE ELEMENT
SUMMER 1952
0
10
20
Miles
KOWON
WONSAN
KOJO
SEA OF JAPAN
YANG-DO
160 Miles
NAN-DO
40 Miles
MO-DO
UNG-DO
YO-DO
SIN-DO
TAE-DO
SO-DO
HWANGTO-DO
CTE 95.23
N
39°

MAP 7

K WHITE

A number of events of major interest occurred during those first difficult weeks following organization of the two offshore island commands. On 19 and 20 February, elements of two North Korean infantry battalions launched an unsuccessful assault against the two Yang-do islands. The combined "action of the island garrison and UN surface forces"[5] repulsed the enemy attempt, which had been planned to gain intelligence and kill as many of the defenders as possible.[6] On the heels of this action, with the first enemy effort to take an east coast island, came an unexpected bonus in the form of a defector. Brigadier General Lee Il, NKPA, came ashore on 21 February at Tae-do "in a stolen sampan with a briefcase full of top secret papers, a head full of top secret plans, and a strong desire to make himself useful."[7] He was rushed immediately to Eighth Army intelligence officers.

The next day command personnel of the west coast TE 95.15 were treated to a surprise, though not so pleasant as the unforeseen defection of the NKPA general. Rear Admiral George C. Dyer, Commander Task Force (CTF) 95, and his staff were engaged in an inspection of the WCIDE islands. While the party was looking over the antiaircraft defenses at Paengyong-do:

> . . . an aircraft of VMA–312 made a pass at the CP, followed closely by a second plane. The second aircraft made a message drop and accidentally released a 500-pound bomb, which landed from 75–100 feet west of the CP, shattering all windows and blowing all the doors off their hinges. Personnel harbored within the CP were thrown to the floor by the concussion, a few sustaining minor cuts and bruises, but no fatalities were incurred . . . Commanding Officer, USS *Bairoko* [the carrier to which VMA–312 was assigned], sent a note of apology to CTE 95.15 and later followed up with material to repair the CP.[8]

In March, CTG 95.1 directed the occupation of Ho-do, barely more than a speck of dirt 4,000 yards south of Sok-to and within 400 yards of the Communist mainland. Despite Colonel Davenport's objection that the proposed action was beyond the defensive mission

---

[5] CinPac Weekly Intel Digest No. 23–52, dtd 6 Jun 52, included as App. 17 to *PacFlt EvalRpt* No. 4, p. 9–110.

[6] First Lieutenant Joseph S. Bartos, Jr., a former All-American football great, also distinguished himself during the Yang-do action. His cool, resourceful, and valiant leadership during the two-day defense earned him the Silver Star Medal. BGen Frank M. Reinecke comments on draft MS, dtd 25 Aug 66.

[7] Field, *NavOps, Korea,* p. 426.

[8] CTE 95.15 ComdD, 1 Feb–31 May 52, p. 8.

of his command and that the proximity of Ho-do to the enemy shore made the island untenable,[9] the task group commander would not rescind the directive. After a detailed reconnaissance by First Lieutenant Wallace E. Jobusch, Colonel Davenport ordered a reinforced Korean Marine Corps platoon to occupy the island. This order was carried out, but during the night of 25–26 March the platoon lost its newly gained objective to a well-coordinated enemy amphibious attack. Not a single Korean Marine survivor could be accounted for at daylight. On 2 April, however, after the enemy force had departed Ho-do, six of the platoon turned up on Sok-to. They had survived by hiding out at Ho-do. None of the others were ever seen again. After the island was overrun, it was not reoccupied by Allied forces.

After this latest offensive action in the west, the Communists made no further attempts to seize any of the islands. U.S. and ROK Marines enjoyed a period of relative freedom from enemy harassment, except for frequent shore battery shelling directed against the east coast islands. For WCIDE command members the quiet island duty was interrupted only occasionally by hostile artillery fire although rumors of imminent enemy landings abounded. On 13 October, however, the enemy bombed Cho-do in the first air attack made against an island garrison since the U.S. Marines had been assigned the west coast island command responsibility. No casualties resulted from this raid. The lull in enemy activity that then ensued enabled island personnel to devote increased efforts towards improvement of their defenses.

Marines instructed, drilled, and conducted tactical exercises for the island forces. Island commanders supervised the construction and improvement of gun pits and other defense installations. At the ECIDE command island, Yo-do, a 2,700-foot airstrip (Briscoe Field) for emergency landings and intelligence flights had been completed by June. Since much of the labor was performed by Koreans, the language barrier sometimes created difficulties. In all these activities the Marines found that they were hampered but not unduly burdened by this problem.

One condition, however, did handicap operations of the island

---

[9] Colonel Davenport later pointed out that the enemy could easily employ high-powered rifles against Ho-do occupants, that resupply posed problems to his command, and that at times the enemy could even walk to Ho-do over the winter ice. Col William K. Davenport ltr to Hd, HistBr, G–3 Div, HQMC, dtd 7 Sep 66.

Marines. This was the supply situation which was prevented from becoming desperate only because the Marines were able to borrow and obtain necessities from other service activities. The inability of the island Marines to draw needed supplies from the responsible U.S. Army agency developed as a result of the slowness of the Marines in approving the task element tables of equipment (T/E),[10] and from insistence of the supplying activity that it would deal only with those units that had approved tables of equipment. The urgency of the situation was alleviated in May when weekly supply flights were begun by the 1st MAW. Even when surface ships did arrive with provisions, Marines frequently discovered that items which had been invoiced were missing.[11] Consumables, especially, had a high rate of disappearance.

Marine Air Operations[12]

Close air support of ground troops remained an almost forgotten mission of Fifth Air Force tactical aircraft. When planes were allotted for close support, both their customary late arrival over the target area and pilot inefficiency left Marine ground commanders less than satisfied.[13] The particular concern of General Jerome, the new 1st MAW commander, was the continuing limited opportunity for his Marines to execute their normal primary mission—close air support

[10] A T/E is a listing of equipment that a unit needs to accomplish its mission. Tables vary according to type of unit and its mission.

[11] Commenting on logistical matters, Colonel Kenneth A. King, who during 1952 commanded first the WCIDE and then 1st CSG, was of the opinion that the main difficulty lay "not in getting requisitions filled, but in getting delivery of what was approved" due to the fact Marines were not assigned to processing of requisitions and delivery of supplies. He had high praise for the concern and assistance of 1st MAW units as well as Captain G. L. G. Evans (RN) of HMS *Ocean* and various other United Kingdom ship captains. Colonel King further commented that "for the benefit of Marines who may have to serve in isolated areas, and I imagine this often prevails in Vietnam today, it cannot be emphasized too strongly that the Marine Corps should be very reluctant to leave the support of any of its elements, no matter how small, to other services or nationalities." Col Kenneth A. King ltr to Hd, HistBr, G–3 Div, HQMC, dtd 24 Aug 66.

[12] Unless otherwise noted, the material for this section has been derived from: *PacFlt EvalRpts* No. 4, Chap. 10; No. 5, Chap. 9; 1st MAW ComdDs, May–Aug 52; MAG–12 ComdDs, Jun, Aug 52; Robert F. Futrell. *The United States Air Force in Korea, 1950–1953* (New York: Duell, Sloan, and Pearce, 1961), hereafter Futrell, *USAF, Korea.*

[13] 1st MAW ComdD, Feb 52, quoted in *PacFlt EvalRpt* No. 4, p. 10–45.

of frontline troops. Although FAF assigned Marine pilots to support the 1st Marine Division whenever possible, the infrequent number of close air support missions performed under the existing sortie limit was beginning to detract from the quality of delivery. General Jerome set out to remedy this unfavorable situation.

Working with General Selden, the Marine wing commander prevailed upon the Air Force to permit close air support training of wing pilots and of forward air controllers with the Marine division. On 19 May, CG, FAF lifted the close air support restriction that he had imposed in front of General Selden's MLR. By agreement between the FAF and the two Marine commanders, Fifth Air Force would permit the scheduling of 12 close air support sorties daily for a one-month period, MAG–12 was given this training mission, to begin on 21 May.[14]

The objective of the CAS program, in addition to providing operational training and practice for Marine ground officers, air controllers, and pilots, was to inflict maximum casualties on Chinese troops and to increase the destruction and damage to their positions. Before assigning a pilot to the actual training flights, MAG–12 sent him on a tour of the front lines to become better familiarized with the topography, the restricted ("no-fly") areas, and probable enemy targets. Air strikes requested by the division went directly to MAG–12. Initially, a limitation of 12 sorties per day was established, but on 17 July—the program having already been extended beyond its original 30-day limit—a new ceiling of 20 daily sorties went into effect. The division was also allowed additional flights above this prescribed daily sortie number when air support was needed to repel a large-scale enemy attack or to assist in a major Marine ground assault.

Almost as soon as the Marines began to derive the benefit of the training program, the flights were terminated by FAF. On 3 August

[14] Two months earlier, FAF had begun "a program for training pilots in close air support techniques. . . . Initially, all training missions for this division were flown by Air Force aircraft." The flights, not in response to specific requests, were assigned by the G–3, I Corps. CG, 1stMarDiv ltr to CG, FMFPac dtd 23 May 52, Subj: CAS sum for pd 1 Jan–30 Apr 52, cited in *PacFlt EvalRpt* No. 4, p. 10–196. These flights ceased just before the ones from MAG–12 began. 1st MarDiv ComdD, May 53, p. 4. A 1st MarDiv staff officer, who had observed the frequency of General Jerome's visits to the division CP to discuss the new close air support training program, has credited the two Marine CGs for their "great amount of coordinated personal aggressiveness in bringing this about." Col Robert A. McGill comments on draft MS, Sep 66, hereafter *McGill comments.*

1952, following a complaint by CG, Eighth Army that Marines were Fifth Air Force notified General Jerome that the special program getting a disproportionate share of the close air support sorties, the would end the next day. General Selden was instructed to request air support "in the same manner as other divisions on the Army front."[15] Despite the abrupt termination of the training program, the division had derived substantial benefits from the 12 weeks of Marine-type close air support. "Air attacks were the most useful weapon for dealing with enemy dug-in on the reverse slopes,"[16] according to an official analysis. One regimental commander reported that the 1,000-pound bombs were effective in destroying enemy bunkers and further noted that the strikes had produced good results in the "destruction or damaging of enemy artillery and mortar pieces."[17] Another senior officer commented that air overhead kept the Communists "buttoned up," which permitted Marines greater freedom of movement for tactical and logistical operations.[18]

A second type of Marine close air support aided the mission of Marine infantrymen in western Korea during the summer of 1952. This was controlled radar bombing, which permitted delivery of aviation ordnance at night or under other conditions of limited or poor visibility. The Air Force had introduced the concept into Korea in January 1951, had tested and evaluated it in combat, and shortly thereafter had put it to good use against the Communist spring offensives that year. Based on a concept oriented towards deep support of troops in extended land campaigns, the Air Force system made use of 20-ton vans to house its ground components.[19]

The Marine equipment, on the other hand, was more mobile since it was to be employed close to friendly lines. Referred to as the MPQ-14,[20] the Marine radar bombing system was designed so that

[15] *PacFlt EvalRpt* No. 5, p. 8–54.
[16] *PacFlt EvalRpt* No. 4, p. 9–36.
[17] 1st MarDiv ComdD, Jun 52, p. 2.
[18] Col Russell E. Honsowetz ltr to Hd, HistBr, G–3 Div, HQMC, dtd 14 Sep 66.
[19] As an Air Force spokesman noted, ". . . the AN/MPQ–2 radars introduced into Korea in January 1951 were Strategic Air Command bomb scoring radars and not tactical equipment. This would explain the large vans." Robert C. Futrell, Historian, Hist Studies Br USAF Hist Div, comments on draft MS, dtd 12Oct66. Dr. Futrell authored the definitive unclassified history of Air Force operations in Korea, previously cited as *USAF, Korea*.
[20] These letters indicate first, the type of installation; next, the kind of electronic equipment; and finally, its purpose. In this case, M-mobile ground installation, P-radar, and Q-intended for a combination of purposes. The figure indicates the model number in the developmental history of the equipment.

the largest piece could be put into a one-ton trailer. Major ground items were a generator power supply, a tracking radar, and a computer; the last essential component, an automatic bombing control, was mounted in the aircraft.

Developed and hand built after World War II by Marines under Major Marion C. Dalby at the Naval Air Materiel Test Center, Point Mugu, California, the MPQ-14 was first used in Korea in September 1951. Initially, considerable mechanical difficulty was experienced with radar bombing, which affected the accuracy of the bombs, but later the system became sufficiently reliable to permit bomb drops within one mile of friendly lines. Subsequent use confirmed the tactical precision of the MPQ-14. By the middle of summer 1952, the Marines had obtained Fifth Air Force permission to use radar bombing, controlled by a forward observer on the ground, in a close support role.

Before this policy change took place another one, at a still higher command level, had occurred. On 23 June, FAF planes struck at eight North Korean hydroelectric plants in the central and northwestern part of the country. The attack represented a departure from the intense interdiction of enemy lines of communication (Operation STRANGLE) which, since May 1951, had characterized FAF support operations. The shift came about after a Far East Air Forces study on the effectiveness of the interdiction campaign had concluded, in part, that the program had been indecisive.[21]

For more than a year preceding the 23 June attack, the Fifth Air Force had concentrated its ground support efforts on the disruption of Communist communication lines so that the enemy would be unable "to contain a determined offensive . . . or to mount a sustained offensive himself."[22] During the lifetime of the doctrine, no major offensive had been launched by the enemy, and on this fact was based the claim for success of the interdiction program. Opponents, however, pointed out that despite this maximum FAF air effort, the Communists had built up their strength, including support areas immediately to the rear of their front lines and resupply installations. As the recent UN commander, General

[21] HistDiv, Air Univ, USAF, *United States Air Force Operations in the Korean Conflict, 1 November 1950–30 June 1952,* USAF Hist Study No. 72 (Washington, 1955), p. 159, hereafter USAF, *Ops in Korea,* with appropriate number. The Air Force operations were published in three books, numbered 72, 73, and 127.

[22] Futrell, *USAF, Korea,* pp. 435–436.

Matthew B. Ridgway,[23] told members of the Senate Committee on Armed Services on 21 May 1952, the same month that FAF had begun to shift its air effort away from interdiction, "I think that the hostile forces opposing the Eighth Army . . . have a substantially greater offensive potential than at any time in the past. . . ."[24]

A number of factors contributed to the reduced emphasis on the interdiction strategy. Three, however, appear to have most influenced the inauguration of Operation PRESSURE, the name given the new policy of concentrating aerial attacks on major industrial targets considered of greatest value to the North Korean economy. Mounting FAF aircraft losses due to enemy flak (fire from ground-based antiaircraft weapons) and an insufficient number of replacements helped shape the new program. By April 1952 FEAF had received "only 131 replacement aircraft of the types engaged in rail interdiction against the 243 it had lost and the 290 major-damaged aircraft on interdiction sorties."[25] These heavy losses had resulted from the increasing accuracy of Communist antiaircraft ground weapons, a capability Air Force planners had failed to consider sufficiently.[26]

Although significant, this loss factor was not the final consideration in executing PRESSURE attacks against the power plants. More directly responsible were two other recent developments. These were the decision of the new UN commander, General Mark W. Clark, to take forceful action to bring the Communists around to an armistice agreement and a top-level Defense Department change of policy that had removed a major North Korean hydroelectric facility from the restricted bombing list. This was the Suiho plant, fourth largest in the world. Adjacent to the Yalu River, about 75 miles northeast of its mouth, Suiho supplied approximately 25 percent of the electrical power used in nearby northeast China.[27]

Results of the PRESSURE strikes, carried on from 23–27 June, were highly successful. Marine, Navy, and Air Force planes flew 1,654 attack and escort sorties in these raids. Of the 13 target plants attacked during this period, 11 were put out of commission and 2

---

[23] General Mark W. Clark had succeeded Ridgway as UN Commander on 12 May 1952. Ridgway was to take over as the new Supreme Allied Commander, Europe, 1 June, replacing General of the Army Dwight D. Eisenhower, who was returning to the United States.

[24] Cited in Futrell, *USAF, Korea,* p. 435.

[25] USAF, *Ops in Korea,* No. 72, p. 156.

[26] Futrell, *USAF, Korea,* pp. 436–437.

[27] *Ibid.*, pp. 452–453 and Cagle and Manson, *Sea War, Korea,* pp. 443–445.

others were presumably destroyed. North Korea was almost blacked out for two weeks. Chinese and Russian experts were rushed to North Korea to lend a hand in restoration. The hydroelectric strikes marked the first time that Marine, Navy, and Air Force pilots had flown a combined mission in Korea. The 23 June strike, moreover, was of particular significance to 1st MAW since it was also the first time that MAGs–12 and –33 were assigned group strikes at specific adjacent targets at the same time.

Led by Colonel Robert E. Galer, the new MAG–12 commander since 25 May, group pilots struck and leveled the single power complex, Chosin 3, in the 23–24 June runs. Colonel John P. Condon, who had taken over MAG–33 on 24 May, put 43 jets from VMFs–311 and –115 into the air during the two-day mission. The first time that its F9Fs had ever been massed for a strike of this type, the MAG–33 jets similarly destroyed the Chosin 4 plant, 11 miles northwest of Hamhung.

Although the jets carried a smaller payload than the Corsairs and ADs of MAG–12 (approximately 37 gross tons to more than 150 tons), the extremely precise bombing record made by the Grumman Panther jet pilots forever put to rest the doubts about jet accuracy that had been held by some in 1st MAW. As the group commander later recalled, "The capability of jet strike aircraft for extremely accurate bombing, an item of open discussion prior to this time, was never questioned in the 1st Marine Aircraft Wing after this mission."[28] Another gratifying result was that flight personnel on all of the 150 Marine aircraft returned safely. In fact, of the total 1,645 FAF sorties, only 2 aircraft were downed; rescue aircraft successfully picked up these two pilots, both U.S. Navy officers.

It was the high probability of being rescued, if forced to abandon their aircraft, that not only eased the minds of pilots on missions north of the 38th Parallel but also permitted the fliers a greater degree of success. As the MAG–12 commander, Colonel Galer, who was shortly to escape imminent capture by the enemy, later declared, "I do know that every pilot flying in this theatre should have the highest possible morale with the knowledge that so many are ready and willing to risk so much to get them."[29]

A Medal of Honor holder from World War II, Colonel Galer

---

[28] MajGen John P. Condon ltr to Hd, HistBr, G–3 Div, HQMC, dtd 1 Oct 66.
[29] MAG–12 ComdD, Aug 52, p. D–4.

was leading a flight of 31 aircraft on 5 August. His objective was the supply area and tungsten mines in the mountainous northeastern part of North Korea, just below the 39th Parallel and 35 miles southwest of Wonsan. After several hits had killed his engine, the MAG–12 commander, preparing to parachute, climbed out over the side of his plane, but found that he had one foot stuck inside the cockpit, probably on the shoulder straps or the loop of the belt. He then pulled himself partially back towards the cockpit, freed his foot with a vigorous kick, cleared the plane, and headed in spread-eagle fashion towards the ground. Almost immediately the plane, falling in a nose dive, caught the descending pilot on the shoulder and pulled him into a spin. Colonel Galer recovered in time, however, to pull the ripcord and thus ease his impact onto enemy terrain. He landed within ten feet of his crashed AU.[30]

"Immediately upon getting free of the chute, I ran as rapidly as possible, staying low, down through a corn field."[31] At the end of the field, the Marine aviator paused momentarily to survey the terrain for an escape route. Spotting a dry stream bed nearby, Colonel Galer dashed toward it and quickly but cautiously moved up it some 100 yards. Then he halted to put into operation a small survivor radio to report his position. The message was received by the rescue air patrol orbiting overhead which relayed the information to pickup aircraft. The patrol advised the downed pilot that a rescue helicopter had already departed for the crash area.

Before breaking radio contact, Colonel Galer told the air patrol his planned movements in order to facilitate pickup. He then quickly left the area which was located too near the crashed aircraft for a rescue attempt. Evading detection by enemy soldiers and curious teenagers moving towards the wreckage, the Marine worked his way to higher ground, keeping the air patrol advised of his changing position. By 1845, a search of the area was underway. Of the events that followed: Colonel Galer wrote:

> At 1908 I heard the helicopter go down the next valley and saw it disappear. I called, told them to make a 180-degree turn since I was in the valley to the southwest and on the north slope. I did not get an answer but soon the helicopter came through a saddle in the ridge. . . .

---

[30] The AU is the attack version of the Marines' famed World War II fighter, the F4U Corsair.

[31] MAG–12, ComdD, Aug 52, p. D–2.

> I immediately let the red smoke (day flare) go, and came out of the bushes . . . calling the helicopter on the radio also. They apparently saw me immediately and came over and hovered. The mechanic leaned out and swung the hoisting sling back and forth. . . . Finally, I grabbed it and got in . . . and the pilot took off. . . . The mechanic pulled me up and into the helicopter as we crossed the valley.[32]

The colonel was not yet out of the woods. The trip to a rescue ship at Wonsan was marked by intermittent bursts of enemy antiaircraft fire. On one occasion the chopper was hit hard enough to spin it completely around. As the rescue craft neared the coast patches of fog added to the hazards of night flying. About this time the warning light indicating low fuel supply came on but "the pilot gambled on making the sea at the risk of having to autorotate through the overcast into the mountains."[33] It was a correct decision. The fuel lasted until the helicopter landed on the rescue vessel. It was then 2100.

Quite naturally the episode brought forth high praise for the rescue system, and particularly for those individuals whose skills, initiative, and courage made downed crew rescues of this type possible. But Colonel Galer also saw some weaknesses. He pointed out that rescue helicopter pilots should be kept up to date on changing enemy flak positions. The Marine group commander also stressed the need for rescue helicopters to establish and maintain a minimum safe fuel level which would depend largely upon the position of the downed aircraft. One final suggestion, not about the system but the aircraft itself, was that fixed-wing aircraft have ejection-type seats. Remembering his own difficulties, the MAG–12 commander further cautioned pilots to be certain they were free of all straps and cords before bailing out.

In addition to attack missions by tactical aircraft and rescue work by its helicopters, the Marine wing was also responsible for providing antiaircraft defense. It was not until July 1951, 13 months after the NKPA invasion of South Korea, that a formal air defense had been established for the country. Fifth Air Force was given the command responsibility of coordinating the aerial defense net for South Korea and its adjacent sea frontiers. In mid-November 1951, the FAF commander had revised the defensive system, dividing his

---

[32] *Ibid.*, p. D–3.
[33] *Ibid.*, p. D–4.

area into a northern and southern sector, at a point exactly halfway between the 36th and 37th Parallels.

FAF commanded the northern air defense sector while the southern sector became the responsibility of CG, 1st MAW. In turn, these two sectors were further divided into subsectors. Each of these, through a tactical air direction center (TADC), maintained radar surveillance of its assigned area and performed plotting and identification functions. Each subsector was charged with being "directly responsible for sector air defense."[34]

Although the 1st MAW commander had been designated as the Air Defense Commander, Southern Sector, Korea, he was not actually given the means to carry out this responsibility. He still did not have command over his tactical squadrons, nor could he exercise control over operations of his tactical air coordination center (TACC) or TADC.[35] Moreover, his southern sector could not originate practice air warning messages. The wing commander had to obtain permission from JOC before he could begin practice intercepts for training his radar intercept controllers.

Several other deficiencies existed in the air defense system that the 1st Marine Aircraft Wing had inherited. There were no ground antiaircraft weapons at the Marine fields until a .50 caliber automatic weapons battery was detached from the 1st 90mm Antiaircraft Artillery Battalion, FMF, early in 1952 and sent to K–3, the home field of MAG–33. Other inadequacies were deficient equipment—a search radar limited to 30 miles out and 20,000 feet up—and lack of an interceptor aircraft capable of rising to meet the faster swept-wing jets the enemy was employing. Airfields housing Marine air groups did not have revetments for either aircraft parking areas or ordnance dumps.

Not all of these weaknesses were acquired with assumption of the air defense mission. There had been a general lack of concern about air defense throughout South Korea. This attitude had resulted from the air supremacy which the Fifth Air Force had quickly established. Camouflage was seldom practiced. Dispersal of aircraft, supply dumps, and servicing facilities was employed only rarely.

---

[34] Futrell, *USAF, Korea,* p. 616.

[35] TACC is the senior agency for controlling all tactical aircraft and air warning functions; the TADC performs similar functions in an area controlled by the TACC. JCS, *Dictionary of United States Military Terms for Joint Usage* (Short title: JD), JCS Pub. 1 (Washington, 1964), p. 141, hereafter *JCS, JD.*

In fact, at K–6, there was not sufficient land to properly scatter installations and aircraft.

Defense of the southern sector was commanded from K–3 (Pohang), the site of the TACC (Major Fred A. Steel). Marine Ground Control Intercept Squadron 1 (MGCIS–1) was set up on the west coast at K–8 and MGCIS–3 (Lieutenant Colonel Owen M. Hines), on the east coast, near Pohang. Each of these intercept units had an early warning detachment operating off the mainland. Antiaircraft artillery was provided by the 90mm AAA battalion, which was controlled, however, by EUSAK. The 1st MAW commander specified a ready alert status for two aircraft during daylight hours. Just before sunrise and sunset, four planes were put on strip alert. Aircraft for night alert were provided by VMF(N)–513 until April, when the requirement was withdrawn. By 30 June 1952, 1st MAW air defense operations had destroyed a total of five enemy planes. The F7F night fighters flown by VMF(N)–513, moreover, had frequently been scrambled to intercept hostile night intruders that had penetrated into the Seoul area, or northern sector.

This low kill rate did little to atone for the steadily increasing number of Marine aircraft lost to enemy flak. Although the number of friendly planes destroyed or damaged in air-to-air combat during the latter half of Korean hostilities diminished sharply compared to the early period, losses due to ground fire were reaching alarming proportions in early and mid-1952. In May 1952 Navy and Marine air losses to enemy action were twice the total for April, and the June figure was even higher. By June, the Communists had massed more than half of their antiaircraft artillery along communication routes that FAF struck nearly every day.

Remedial action was soon taken. Stress on flak evasion was emphasized in pilot briefings and debriefings. The MAG–33 intelligence section came up with a program that attempted to reduce losses by a detailed analysis of flak information. The originator of this system, First Lieutenant Kenneth S. Foley, based his method on:

> . . . photo interpretation of an up-to-date flak map, scale 1:50,000, and an intelligent utilization of flak reports disseminated by the 67th Tactical Reconnaissance Squadron of the 5th Air Force. Frequent briefings were given to each squadron on the enemy AA capabilities. Elaborate overlays were drawn up and displayed. Target maps, clearly showing AA positions and flak clocks [danger areas], were given to flight leaders to aid them in evading known AA guns in their target area. Through flak analysis, the

safest route to the target area was determined and an actual attack and retirement route was suggested. These recommendations appeared in a flak summary presented at each combat briefing.[36]

Other measures attempted to reduce mounting losses of personnel and aircraft. In all Marine air units, evasion and escape tactics were stressed. In addition to the FAF de-emphasis on interdiction of communication routes that had come about, in part, due to heavy aircraft losses, Fifth Air Force decreed that beginning 3 June, "with the exception of the AD and F4U aircraft [1st MAW types] only one run will be made for each type of external ordnance carried and no strafing runs will be made."[37] CTF 77 ordered that in all attack runs, aircraft would pull out by the 3,000-foot altitude level. The Marines, combining their air and ground efforts, came up with a positive program of their own. It was to become the first known instance of Marine ground in support of Marine air.

Although the originator of the idea cannot be positively identified, the time that artillery flak suppression firing was first employed can be traced back to late 1951, when the division was still in East Korea.[38] It was not until June 1952, however, that a published procedure for conducting flak suppression firing appeared in Marine division records. That same month another type of flak suppression, this by an aircraft, was utilized by the 1st Marines, commanded at the time by Colonel Walter N. Flournoy. The procedure called for the FAC [forward air controller] to relay gun positions to friendly strike planes which temporarily diverted their attack to silence the located gun. Although the method "worked with good results,"[39] it was not destined to become the system adopted by the Marines.

The more frequently used flak suppression called for artillery to

---

[36] VMF(N)–513 ComdD, Jun 52, App II, p. 5. Mention of a flak analysis program first appeared in the March 1952 records of MAG–33. Aircraft losses on interdiction strikes (the program was not applicable to CAS missions) dropped for the next several months. When Lieutenant Foley transferred to the night squadron, he took his system with him and had it put into operation there. LtCol Kenneth S. Foley interv by HistBr, G–3 Div, HQMC, dtd 24 Mar 66.

[37] FAF CbtOps Notam No. 6–10.1 cited in App. 9, *PacFlt EvalRpt* No. 4, Chap. 10, p. 10–199.

[38] *Pala comments; Nihart comments.* Both of these officers, the former artillery, the latter infantry, recall flak suppression firing late in 1951 or early in 1952 when the division was on the eastern front. Colonel Nihart pointed out, in addition, that "such expedients and new tactics went on for some time before getting into the regimental commander's reports."

[39] 1stMar ComdD, Jun 52, p. 2.

fire on hostile gun positions that could impede the success of a friendly close air support strike. Several Marine officers appear to have had a major role in the development and employment of this technique. Among them were Brigadier General Frank H. Lamson-Scribner, Assistant Commanding General, 1st MAW; Colonel Henderson, the 11th Marines commander; and Lieutenant Colonel Gerald T. Armitage, 3/1 commander.

The 1st Marines battalion commander explained how the system operated in late spring 1952:

> I was in an outpost watching an air strike. I asked Captain Shoden [John C., the battalion forward air controller] to work out some idea of flak suppression. Shoden, G–2, and others worked two or three weeks to complete the first plot of antiaircraft positions. My idea was to have a plane start a run and then pull up before finishing the dive. The enemy antiaircraft gunners could not tell that the pilot was pulling out at an extremely high level. The batteries would fire and Marine observers would plot their positions from their fires. Then, the Marine artillery would lay a heavy barrage on these positions.[40]

While observing an air strike from the Marine division sector, General Lamson-Scribner noted that prior to the strike there had been no preparatory firing on enemy antiaircraft artillery positions. After the strike he discussed this matter with General Selden, who "directed me to discuss with his chief of staff what I had observed and my suggestions that division firepower for 'flak suppression' be coordinated with air strikes."[41] The upshot of this was that the division chief of staff suggested that the 11th Marines regimental commander and his staff members develop an SOP[42] for using artillery flak suppression fires in support of close air support strikes. It was believed that proper utilization of these fires would reduce aircraft losses and further increase the opportunity for a successful close air support mission by destruction of enemy antiaircraft weapons.[43]

---

[40] LtCol Gerald T. Armitage interv by HistBr, G–3 Div, HQMC, dtd 15 Aug 61.

[41] MajGen Frank H. Lamson-Scribner ltr to Hd, HistBr, G–3 Div, HQMC, dtd 12 Oct 66.

[42] An SOP, standing operating procedure, is a set of instructions for conducting operations that lend themselves to established procedures. *JCS, JD,* p. 133.

[43] With respect to the effect of enemy fire on attack aircraft, the CO, MAG–33 later commented that "Antiaircraft artillery has a direct deterioration effect on pilot accuracy, particularly with regard to care in getting on target and doing a precise job." CO, MAG–33 ltr to CG, 1st MAW, dtd 25 Jul 52, quoted in *PacFlt EvalRpt* No. 5, p. 9–76.

On 30 June 1952, the 11th Marines published the SOP. Since the objective was to prevent enemy fire from interfering with friendly strike planes, the key to the entire procedure was the precise coordination of artillery fire with the delivery of aircraft ordnance. As Colonel Henderson described the system:

> When the infantry regiment received word of an air strike, the air liaison officer plotted on the map . . . the target of the strike, the orbit point, the direction of approach, and the altitude . . . and direction of pullout. Then the artillery liaison officer, by looking at the map, could determine which of the Chinese positions could bring effective fire on the strike aircraft. The artillery battalion had prearranged code names and numbers for every antiaircraft position. All the artillery liaison officer had to do was pick up the phone and tell the F.D.C. [fire direction center] 'flak suppression' and read off what targets he wanted covered.
>
> These fires were then delivered on the request of a forward observer who was with the forward air controller . . . When there was a forward air controller up in the front lines controlling the strike, we would put a forward observer with him. When the planes were . . . ready to go, the F.O. [forward observer] got the word 'Batteries laid and loaded,' and he would tell them to fire. The minute the FO would get the word, 'On the way,' the forward air controller would tell the planes to start their run. As a result, we had cases where the planes were in their bombing run within 30 seconds after the flak suppression was fired, which meant that they were in on the target while the positions were still neutralized. The question of control and split second timing is of exceeding importance because the aircraft are going 300 to 400 miles an hour. . . .[44]

Early in the program the MAG–12 commander reported that although the flak suppression procedure was not flawless, it was proving "very capable and workable."[45] An indication of the success of 1st Marine Division pioneering efforts in flak suppression is seen in the fact that shortly after it was put into operation "there was a steady stream of visitors to the 11th Marines CP to find out what [it was] and how we were doing it and to get copies of our SOP."[46] The procedure was eventually adopted by other Eighth Army units.

Marine air losses from hostile ground fire during CAS strikes immediately began to drop from the June peak and never again reached this level. In 124 close support sorties flown by 1st MAW on 13 August, not one plane was shot down and only four received

---

[44] *Henderson ltr II.*

[45] CO, MAG–12 Spdltr to CG, 1st MAW, dtd 2 Jul 52, Subj: Comments on 11th Mar Flak Suppression SOP, cited in *PacFlt EvalRpt,* No. 5, Chap. 9, p. 9–78.

[46] *Henderson ltr II.*

minor damage from enemy flak. Although there were some complaints as to execution of the flak suppression program these would be corrected, in the main, by a revised procedure which the 11th Marines would undertake in the winter of 1952.

Spring 1952 on JAMESTOWN[47]

Earlier in the year the Marines had revised their estimate of enemy capabilities after the lengthening of the division MLR by I Corps and the subsequent heavy enemy attack. The re-evaluation placed the most likely course of Chinese action as defending their present positions with the 21 infantry battalions assigned and also cautioned that the Communists could mount a limited objective attack at any time of their choosing. Division intelligence estimated that the Chinese could muster up to "57 infantry battalions supported by 12 artillery battalions and 40 tanks and/or self-propelled guns" for a thrust into the Marine sector.[48]

The enemy, however, showed little disposition for any concerted ground attack during the remainder of April. But before the month ended, Marines, in conjunction with other I Corps divisions, had deluged the enemy with artillery and tank fire in Operation CLOBBER. The purpose of this shoot was to inflict maximum casualties and damage by employment of the element of tactical surprise. The reinforced 11th Marines, augmented for this occasion by Company D, 1st Tank Battalion and nine of the battalion's 105mm howitzer and flame tanks, blasted Chinese CPs, bivouac areas, artillery and mortar positions, and observation posts. Marine frontline regiments joined in with their organic mortars. Since most of the firing took place at night when results were unobserved, no estimate could be made as to the effect of the operation on the enemy.

A new Marine artillery tactic about this time was the counter-counterbattery program instituted by the 11th Marines. The regiment had developed this technique to counter superior enemy artillery strength. This situation, as well as the fact that I Corps artillery

[47] Unless otherwise noted, the material in this section is derived from: *PacFlt EvalRpts* No. 4, Chap. 9, No. 5, Chap. 8; 1stMarDiv ComdDs, Apr–Jun 52; 5thMar ComdDs, Apr–Jun 52; 7thMar ComdD, Jun 52; 11thMar ComdDs, Apr–May 52; 1/5 ComdD May 52; 1/7, 2/7 ComdDs, May 52.
[48] 1stMarDiv ComdD, Apr 52, p. 1.

available to the division was considered inadequate for counter-battery support, led the Marine division to adopt the new program in May 1952. One provision required a battery in each battalion to select counter-counterbattery positions and occupy them for 24 consecutive hours each week. Another proviso of the program was the selection by each battalion of 10 roving gun positions that were to be occupied by a single weapon rotated to each place at least once weekly. By these tactics, the artillery regiment hoped not only to mislead the Chinese in their estimate of the strength and location of Marine artillery but also to dilute enemy counterbattery intelligence by causing him to fire into areas just vacated by friendly guns. "The effectiveness of the program was demonstrated on numerous occasions when the enemy fired counterbattery into unoccupied positions."[49] An added advantage was that of providing deeper supporting fires on target areas.[50]

Still another concept regarding the employment of artillery developed during the early days of the JAMESTOWN defense. The 11th Marines had advised the infantry regiments that it could effectively fire on enemy troops attacking friendly positions if the Marines had overhead cover. The idea was to use variable time (VT) fuzes[51] with the standard high explosive (HE) shells. Artillery battalions supporting the frontline regiments registered on positions occupied frequently by patrols going forward from JAMESTOWN.

According to the recollections of veteran artillery and infantrymen in the division, the first occasion that pre-planned artillery fire was placed on friendly positions occurred in May 1952.[52] The episode involved a 2/7 platoon patrol that late on 18 May was ordered to return to the MLR from an outpost on the former OPLR. Operating forward of the center regimental sector,[53] the platoon commander, Second Lieutenant Theodore H. Watson, directed that two of the three Marine squads return to the MLR. The remaining unit, surrounded by about 50 Chinese, engaged them in a brisk fire fight.

---

[49] *PacFlt EvalRpt* No. 5, p. 8–51.

[50] LtCol Bruce F. Hillam comments on draft MS, dtd 31 Aug 66.

[51] A type of proximity fuze, the V.T. depends upon an external source, such as an electronic signal, rather than the force of ground impact, to detonate the shell at a predetermined height over the target.

[52] 1stMarDiv ComdD, May 52, p. 4.

[53] The 7th Marines advanced to the line to relieve the 5th Marines in the center sector on 11 May.

When the artillery fired to seal off the enemy and box-in the defensive position failed to discourage the hostile force, Lieutenant Watson ordered his men into the shelter of two nearby bunkers. He then requested the artillery to place VT directly over his positions.[54] The volleys of overhead fire and effective Marine small arms fire then forced the enemy to call off his assault. Although the exact number of Chinese casualties could not be determined, the new fire technique fully accomplished its purpose—repelling the enemy force.

Initiating the infantry action in May was the 1st KMC Regiment, holding the division left flank, with its 2d and 1st Battalions on line. At dusk on 3 May a platoon-size raiding party, under Second Lieutenant Kim Young Ha, left an outpost forward of the 1st Battalion line on a prisoner-taking mission and headed for the objective, Hill 34, adjacent to the rail line to Kaesong and about a half-mile west of the Sachon River. When the platoon was within approximately 1,000 yards of its goal, a support squad was detached near a trail and stream juncture to ambush any enemy attempting to attack the raiders from the rear.[55] The remainder of the platoon, two assault squads, then continued towards the objective, moving cautiously and halting for an hour because of the bright moonlight.

After midnight the moon disappeared behind the clouds, and the Koreans again emerged. They advanced towards a village immediately south of the objective. After searching a few houses and not finding any enemy, the KMCs started on the last leg to Hill 34. As soon as the objective came into view the raiders deployed for the assault. At 0410 the two squads of Korean Marines charged the knoll, immediately drawing heavy Chinese small arms fire. When the raiders continued their assault, the enemy retreated to his trenchworks and bunkers where he continued to fire on the KMCs. Since it now appeared to the patrol leader that the probability of taking a prisoner was unlikely, he prepared to return to friendly lines. He first arranged for artillery to cover the withdrawal of the patrol, and then broke off the 18-minute fire fight, taking his only casualty,

---

[54] The artillery regiment had earlier developed the "box-me-in" fires for outpost defense. If under heavy attack the outpost could call for these preplanned close-in fires that completely surrounded the position. In event of radio or wire communication failures, the outpost could call for "box-me-in" or "Fire VT on my position" by signal flare or other pyrotechnic device. *Henderson ltr II.*

[55] This support squad itself was later ambushed. The heavy casualties it received prevented its further participation in the raid. KMC Regt UnitRpt 53, dtd 4 May 52.

a wounded rifleman, with him. The KMCs counted 12 enemy dead. No prisoners were taken. In the preliminary action, the support squad had also suffered three killed and seven wounded.

As the KMC raiders were making their way back to the MLR, a combat patrol from 1/5, the reserve battalion of the 5th Marines, prepared to move out. This patrol was one of many dispatched by the battalion during the first week of May in accordance with its mission of patrolling in front of the OPLR, between the MLR and the OPLR, and throughout the regimental sector. On this occasion, the patrol was to occupy the high ground south of former Outpost 3, which had become the focal point of activity in the center sector.[56] When used as a base of fire, this ground provided a position from which automatic weapons could readily cover enemy lines or tie in with adjacent friendly defenses. In addition, the 1/5 patrol was to drop off friendly snipers to cover the former OPLR position, to maintain surveillance, and to ascertain to what extent the Chinese were developing the outpost. The task went to a Company A platoon, which the unit commander, First Lieutenant Ernest S. Lee, reinforced with light and heavy machine guns.

At sunup the Marines crossed line JAMESTOWN and before 0900 had reached the high ground they were to occupy. Here the patrol leader set up his base of fire, then pushed on with the rest of his men to the outpost, receiving occasional mortar fire before reaching the old position. While organizing his men at the objective, Lieutenant Lee received word by radio that the Chinese were preparing to attack. Almost immediately, intense shelling struck the forward slope of the hill. A Marine aerial observer (AO) detected 60–70 Chinese advancing from the next hill, some 800 yards to the front of the Marines. The AO also reported that the enemy was firing mortars towards OP 3.

Shortly thereafter the Chinese fire ceased. Moments before it lifted, the patrol received a second warning that an enemy attack was imminent. Even as this message was being received, about 30 Chinese rushed the patrol. The Marines immediately took the hostile assault force under fire, killing 14 CCF with well-placed small arms

---

[56] This position, the site of the mid-April battle, along with several others had been abandoned when the division withdrew its OPLR late in April. Infantry regiments dispatched frequent patrols in an attempt to discourage the enemy's incorporating the hill into his own OPLR.

fire. Overhead, four 1947-vintage Marine Corsair fighters (F4U-4Bs) struck at troublesome mortar positions previously located by the AO. At 1330 another aerial strike against Chinese mortars and enemy positions on the hill north of OP 3 was executed. These two air missions were credited with destroying six mortars, damaging two others, and wrecking seven personnel bunkers. During the second strike the 1/5 patrol began its withdrawal.

On two occasions during the patrol's return to its base the enemy attempted to ambush it. Each time the attempt was thwarted, once by the patrol itself and the second time, with the help of friendly artillery. On the way back several loud explosions suddenly halted the patrol. Investigation revealed that the Marines, carrying their casualties of one dead and four wounded, had inadvertently stumbled onto a path not cleared of mines. Two members of the stretcher bearer detail were killed and three others wounded by the AP (antipersonnel) mines that had not been charted on friendly maps by the Marines' predecessors in the defense sector. A mine clearance team promptly disposed of the danger. With the aid of fires from a 2/5 patrol on the nose of a nearby hill, the 1/5 platoon was able to break contact. After pulling back several hundred yards, the patrol reached a forward medical aid station where jeeps picked up the more seriously wounded and took them to helicopters, which completed the evacuation. Patrol members reported 27 known enemy dead, including one that had been propelled into the air by a direct hit from an artillery round.

The next major Marine ground action soon involved the same Company A platoon, but this time as part of a larger force. Colonel Culhane, the regimental commander, directed his 1st Battalion to launch a new raid on the Outpost 3 area in an attempt to oust the Chinese and thereby deny the enemy use of the critical terrain. Inflicting casualties and capturing prisoners were additional tasks assigned. On 8 May Lieutenant Colonel Nihart issued Operation Order 12–52, calling for 1/5 to seize a series of three intermediate objectives (S, V and X) en route to OP 3 (Y). (See Map 8.) The combat patrol, reinforced by regimental elements, less Company B, was to be prepared to move north of OP 3 to occupy the next hill mass (Z), if necessary.

Operational plans called for Lieutenant Lee's Company A to do most of the leg work as the assault unit. Captain Leland Graham's

OBJECTIVES FOR 1/5 ATTACK
9 MAY 1952
1000 500 0 1000 Meters
1000 500 0 1000 Yards
LINE OF DEPARTURE
MLR
Z
Y
X
V
S
200
150
100
50

MAP 8

K. White

Company C, the diversionary force, was to make a feint against Hill 67, an enemy position southwest of OP 3, and to neutralize it by fire. Weapons Company, under First Lieutenant Ross L. Tipps, in support of the Company A force, was to set up a base of fire at a designated position (T), southeast of OP 3. Artillery support was to be furnished by 1/11, 4/11, and the attached 4.5-inch Rocket Battery. A section of regimental 4.2-inch mortars was also assigned. One platoon of Company B tanks was to assist the assault force by firing both on designated positions and targets of opportunity. Close air support flights were to be on station at two periods during the 9 May daylight operation.

In the early morning hours, under cover of darkness, all units moved into position. At 0430 the 1st Platoon of Company A crossed the line of departure heading for Objective S, a small ridge south and west of OP 3. The 2d Platoon followed and moved out on the right, while the 3d Platoon covered the rear. This hill, lightly defended, was quickly overrun by the Marines. The 1st Platoon then turned northeast towards the four peaks (designated as V, X, Y, and Z), its main objectives. These four positions were all situated at approximately the same elevation, 450 feet. A distance of some 1,300 yards separated the first and fourth hills in the north-south ridgeline.

As the 1/5 platoon neared Objective V, friendly rockets lashed the crest of the hill, which was held by a reinforced enemy platoon in mutually supporting fighting holes. Assisted by this fire, Marine two-man teams with rifles and grenades assaulted the fighting holes occupied by the Chinese. As the Marines proceeded to clear the objective, half of the Chinese were forced to retreat to safer ground. Marines estimated that 15 enemy were killed and a like number wounded. By this time, three hours after setting out on the raid, the platoon had seized one prisoner and sustained five wounded.

While reorganizing for the attack against Objectives X and Y, the 5th Marines patrol came under a heavy artillery and mortar barrage that killed one Marine and wounded three others. As the main body of the assault force advanced towards Objective X to support the attack, the lead elements of the company headed for OP 3. Throughout this maneuver, the company remained under heavy artillery fire.

Proceeding along the eastern slope of the ridgeline to assault knobs X and Y, the platoon had a good view of the effectiveness of their friendly supporting artillery fire. In fact, the combined rocket, howitzer, mortar, tank, and machine gun fire threw up so much dust that at times it restricted the vision of the Marine assault team. As platoon members neared the summit of Objective X they encountered a heavy stream of defending fire. A strong counterattack from the front and left flank assailed the 1st Platoon, but the Marines repulsed the enemy with accurate small arms fire, killing six CCF. Infiltrators then attempted to envelop the Marine platoon and isolate it from the rest of the Company A assault force. Successive waves of Chinese, employing a wedge formation, tried to overrun the main body of the assault force. In repulsing this latest counterattack, Company A killed 12 and wounded 5 enemy.

Quickly sizing up the situation, the company commander ordered the 1st Platoon to rejoin the rest of the assault force. As the platoon began to pull back at 1435 the Chinese blanketed the route with a heavy barrage, firing "over four hundred rounds in a five minute period."[57] This intense shelling took the lives of three Marines, wounded a number of others, and halted the assault force just short of its final goal. Even though the Chinese had been driven from the three intermediate objectives, the devastating enemy mortar and artillery fire made the Marine position untenable. A third of the platoon moved back to Objective V; the rest worked their way along a route east of that objective. While the rest of Company A and Weapons Company elements occupied Hill T, the diversionary force, Company C, reinforced by other Weapons Company personnel, had remained at a strongpoint not far from Objective S. All supporting ground weapons assisted in the withdrawal. In addition to lending direct fire support, Marine tanks brought forward emergency supplies and evacuated casualties. By 1730, the assault force had returned to friendly lines, followed shortly by the rest of the battalion.

Although the battalion failed to seize and hold all of its objectives, that part of the mission calling for inflicting casualties and taking

---

[57] 1/5 ComdD, May 52, p. 10.

prisoners had been successfully executed.[58] Marines counted 35 enemy dead, 53 wounded, and 1 POW, and estimated that an additional 70 CCF had been killed and 105 wounded. Seven Marines were killed and 66 wounded in the action described by some observers as "the largest offensive effort the 1st Marine Division [has] made since last September."[59] The battalion fire support was well controlled and coordinated from an observation post on the MLR. Five air strikes, including one MPQ-14 mission, were credited with destroying three artillery pieces and an equal number of mortars, damaging two other mortars, and demolishing six personnel bunkers.

As the regiment noted, the earlier withdrawal of the OPLR had "altered to a considerable extent the tactics employed in this area. This is especially apparent in the number of patrol contacts close to the MLR and displayed the eagerness of the enemy to move in on any ground not held by friendly forces."[60] At the same time the increased number of troops made available for the MLR defense considerably strengthened the JAMESTOWN Line itself. Sector responsibility changed on 11 May. Colonel Russell E. Honsowetz' 7th Marines relieved the 5th Marines in the center regimental sector, with 2/7 and 1/7 occupying the left and right battalion positions, respectively.

When it took over the peace corridor sector the 7th Marines also assumed the responsibility for emergency rescue of the Allied truce delegates at Panmunjom.[61] The regiment advanced a mile nearer the objective when it moved the pick-up force's assembly area to within 400 yards of the line of departure. The 7th Marines also replaced the tanks in the force with M–39 personnel carriers, a U.S. Army-developed tracked vehicle similar in appearance to the Marine amphibian tractor. Another vehicle the 7th Marines retained in its task force was a medium tank equipped with additional radios. This armored communication and control vehicle was used as a radio

---

[58] Lieutenant Colonel Nihart believed that the heavy enemy shelling, which had caused the early retirement of his battalion, had been possible either because Chinese mortar and artillery positions were so well camouflaged that intelligence had not located them or else so well protected that UNC counterbattery fire had failed to destroy them. *Nihart comments.*

[59] 5thMar ComdD, May 52, p. 9.

[60] *Ibid.*, p. 1.

[61] This force and its mission at various times were known as "Task Force Jig" or "Operation Snatch."

relay station on the MLR to assist in liaison between moving infantry and tank units. Marine riflemen dubbed this command tank the porcupine, to describe the effect of many bristling antennas sticking out from its top. While the Marine division right sector, occupied by the 1st Marines, remained relatively quiet during the spring months on JAMESTOWN, the 7th Marines in the center MLR would shortly be involved in the division's major ground action in late May.

As part of the active defense of its JAMESTOWN line, Lieutenant Colonel Daughtry, commanding 1/7, issued a directive on 26 May intended to deny to the enemy key terrain remaining on the old OPLR. Operation Plan 16–52 called for an attack to seize two parcels of high ground to the regiment's right front. At the same time, the battalion was to neutralize two Chinese positions west of the main objectives, Hill 104 (Objective 1) and the Tumae-ri Ridge (Objective 2), approximately a half-mile further north. The designated attack force, Captain Earl W. Thompson's Company A, was heavily reinforced. While Company A pursued its mission to the right, a Company C reinforced platoon under Second Lieutenant Howard L. Siers would conduct a feint on a pair of enemy positions to the left. Support for the operation would come from 2/11, two tank platoons, and from air, which was to be on call.

H-Hour was set for 0300 on 28 May. Attack and diversionary forces on schedule crossed the line of departure, a half-mile north of the MLR. Captain Thompson's main force advanced nearly to the base of Hill 104 before the Chinese, in estimated reinforced platoon strength, began to counterattack. The fight came to an abrupt end when Second Lieutenant John J. Donahue led his platoon to the top with bayonets fixed.[62] As the Marines dug in they came under heavy mortar and artillery fire from CCF strongholds to the north. On the left, meanwhile, Lieutenant Siers had received orders to seize the closer of his two objectives, former OPLR 5, instead of merely placing suppressive fire on it.

Moving forward from its base of fire, the platoon soon established contact with the enemy. At 0554 the platoon began its attack on the objective. Despite the close-in, hand-to-hand fighting, when it became apparent the assault could not be stopped the enemy gave

[62] Maj Kenneth A. Seal comments on draft MS, dtd Oct 66. At the time of this attack, Lieutenant Seal commanded the 2d Platoon, A/1/7.

way to Marine persistence in seizing the hill. By 0700 the Company C, 7th Marines platoon had secured its objectives and begun preparations for defense of the positions as well as continued support of the main attack force. Heavy casualties, however, forced Lieutenant Colonel Daughtry to recall the platoon and it returned to the lines by 0930.

Up on Hill 104, Company A, 1/7 faced practically the same situation. Taking Objective 1 had been costly and the advance through withering enemy fire was adding to the casualties. A reinforcing platoon was sent from the MLR to help the company disengage and return to friendly lines. Contact with the enemy was broken shortly after noon. With the aid of air and artillery, the company was able to make its way to the MLR by 1405.

Advancing only as far as it did, the attack, like the one earlier that month, failed to take all the designated objectives. Casualties to the 1/7 Marines were placed at 9 killed[63] and 107 wounded. Most of the latter were evacuated for further treatment. Forty-five of the enemy were counted dead and three wounded. Marines estimated another 40 enemy killed and 40 more wounded.[64] The action resulted in a casualty toll that was the highest to date for any Marine company in western Korea. All three Company A rifle platoon leaders—Second Lieutenants Donahue, Jules E. Gerding, and Kenneth A. Seal—were wounded. This battle also became the occasion for another unwelcomed record—4,053 rounds of enemy incoming, during a 24-hour period.

Following this late May offensive, a brief period of relative calm settled over the MLR. Marine and Chinese units continued the active defense of their respective sectors, with generally only a limited number of contacts. Fire fights between Marine patrols and CCF defenders lasted only a short time and usually ended when artillery

---

[63] Two Marines killed in the action were later posthumously awarded the Medal of Honor. Corporal David B. Champagne, A/1/7, was responsible for saving the lives of the three other members of his fire team. When a grenade fell in their midst, Champagne grabbed it to hurl back to CCF positions. Just as it cleared his hand, the grenade exploded, showering lethal shrapnel into the body of the 19-year-old Rhode Islander. One of the C/1/7 reinforcement Marines, Private First Class John D. Kelly, had conducted a one-man assault against a dug-in Chinese machine gun crew. Though painfully wounded during this encounter, he disposed of the enemy, then reduced a second weapons bunker. While firing point-blank into a third position the brave Marine was fatally wounded. This 1/7 action was the first in the western Korea defense to result in multiple Medal of Honor awards.

[64] 1/7 ComdD, May 52, pp. 17–18.

fire caused the patrol to pull back. Even though this state of affairs remained essentially unchanged through June, several other events that month would affect Marine defense of the westernmost sector in I Corps.

End of the Second Year of War[65]

A second realignment of the Marine-Commonwealth boundary along Line JAMESTOWN was made on 1 June. Part of the rear of the MLR was moved eastward to enable the Marine division to assume full responsibility for a key ridgetop. Prior to this date the hill mass had been divided along its crest, a factor that made it a potential trouble spot for both divisions. On 23 and 24 June, the 7th Marines MLR battalions relocated their positions towards the enemy along JAMESTOWN. This readjustment of the line varied from 1,300 yards in the center of the regimental sector to 400 yards near its right. The additional terrain strengthened the division front by placing the center regiment on improved and more defensible ground.

A week before this MLR change took place, there had been a shift in occupants in its far right sector. Colonel Culhane's 5th Marines replaced the 1st on line, which then went into division reserve. Manning the MLR were 2/5 on the left and 1/5 to the right.

In early June the recently appointed UN commander, General Clark, made his first visit to the 1st Marine Division front. During his briefing, General Selden reviewed the unusual combat difficulties confronting his Marines. In addition to the unfavorable terrain, the division commander noted the special operational restrictions caused by proximity to the truce talk site. Presence of a large number of uncharted minefields created another obstacle. Herculean efforts were required of the Marines to simultaneously man and construct defenses over 35 miles of JAMESTOWN. Adding to Marine problems were the facts that ground units were not receiving sufficient close air

[65] Unless otherwise noted, the material in this section is derived from: *PacFlt EvalRpts* No. 4, Chap. 9, No. 5, Chap. 8; 1stMarDiv ComdD, Jun 52; 5th Mar ComdDs, Apr, Jun 52; 7thMar ComdDs, May–Jun 52; 1/7 ComdD, May 52; KMC Regt Unit Rpt 120, dtd 30 Jun 52.

support and the capabilities of the Chinese were constantly increasing.

Chinese order of battle (OOB) information was fed into the division intelligence network by higher commands, I Corps and EUSAK, and adjacent units, but a large part of the data about Communist forces was produced by the division itself. Frontline units in contact with the enemy, by observation of his activities, supplied the bulk of intelligence about enemy defense tactics, employment of weapons, and combat characteristics. Supporting Marine division units, particularly artillery and armor, fed more facts into the system, mostly through identification of the caliber of enemy shells fired at the Marines. As a result of its missions forward of the line and actions in defense of it, the division reconnaissance company also contributed to the intelligence network. Individual Marines, performing as tactical air observers and artillery air observers, as well as the VMO and HMR pilots, were other important sources readily available to the 1st Marine Division.

G–2 directed the division intelligence effort, including processing of raw material and supplying of updated reports to 1st Division units. The G–2 section also maintained OOB and target identification data on Chinese units and their commanders. Members of the G–2 staff also assisted in interrogation of prisoners of war (POWs), screened the civilians apprehended in unauthorized areas, debriefed Marines exposed to enemy intelligence, and conducted inspections of division internal security. In areas where the 1st Marine Division had only a limited intelligence capability it turned to EUSAK for assistance.

Eighth Army teams augmented the division counterintelligence efforts and provided most of the translation service. In addition, three radio intercept units furnished information to the Marines. The critical importance of this service had been proven during several combat patrols in May when additional information was instantly radioed to a friendly unit under fire.

Other intelligence activities were less beneficial to the Marines. These operations were conducted by Tactical Liaison Officers (TLOs, friendly Koreans trained by U.S. intelligence teams), and members of a Higher Intelligence Detachment (HID), a Korean unit assigned from EUSAK. Both the TLO and HID proved of limited value to the division, due to the generally poor educational background of the agents, their inadequate training, and frequent failure to return from assignments behind enemy lines. Some Marines believed the basic fault in these operatives lay in "an exaggerated opinion of their importance."[66]

Several division intelligence Marines, in conjunction with training and shore party personnel, took part in an informational activity of a different type. These Marines reconnoitered several friendly islands off western Korea to determine their suitability for division landing exercises. The second one inspected, Tokchok-to, 30 miles southwest of Inchon, was selected. By early June planning had progressed to the point where a program had been developed for bimonthly battalion landing team exercises. The KPR maneuver force, appropriately reinforced, was designated as a participating unit. Landings were to employ boat teams, amphibian tractors, and helicopters. The entire program was designed to provide refresher training for Marines in carrying out their primary mission of amphibious assault. By the end of June, 3/5 and 3/1, in turn, had captured Tokchok-to.

Other training concentrated more on the task at hand. Division units in reserve rehearsed tactics for offensive and defensive warfare. Most ground units conducted extensive schooling in both mine and booby trap detection and clearance. Recognizing that patrolling was an important part of a Marine's life on the MLR, the division included in its Noncommissioned Officers' (NCO) Leadership School a thorough indoctrination in patrolling tactics.[67] More than 50 percent of the training at all levels was at night. In addition, an extensive orientation was conducted for newly arrived combat replacements, who could not be committed to action for 72 hours after joining the division.

A week after the division's June replacements landed at Inchon, General Selden's headquarters received a directive that would affect a number of these new Marines. On 10 June CG, EUSAK ordered his corps commanders to make continuous efforts to secure the identification and changes in the enemy order of battle. Two days later I Corps followed the Eighth Army order with a letter of instruction which called for each I Corps division to "prepare plans for launching swift, vigorous, and violent large-scale raids to

---

[66] *PacFlt EvalRpt* No. 4, p. 9–33.

[67] BGen Austin R. Brunelli ltr to Hd, HistBr, G–3 Div, HQMC, dtd 13 Sep 66, hereafter *Brunelli ltr*. The division chief of staff during more than half of 1952, Colonel Brunelli later observed that the "school produced more effective patrolling and . . . contributed to reducing our casualties."

capture prisoners, to gain intelligence, to destroy enemy positions and material and/or strong limited objective attacks to improve and strengthen Line JAMESTOWN."[68] Large scale was defined as an "attacking force limited to battalion or regimental (brigade) size with appropriate armor and artillery support."[69] Divisions were required to submit detailed proposals for future action by 21 June. Marine division plans for limited objective attacks during July by units of the 7th Marines and KMCs were subsequently prepared and forwarded to I Corps.

One operation conducted north of the 2/5 left battalion sector early on 22 June was not, however, in response to this enemy identification mission. Late the previous day, Company G had sent out a 16-man ambush. Before the Marines reached their destination, a small enemy force, itself lying in wait, began to pour a heavy volume of fire on the Marines. At this point the patrol was ordered to pull back. One group of 10 made it back to the MLR; the remaining Marines headed for a nearby combat outpost in friendly hands. Reports to the company revealed one Marine not accounted for. The outpost commander was directed to search the area for the missing Marine. This reconnaissance by a fire team failed, but a reinforced squad sent out later brought back the body of the Marine who had been killed by Chinese artillery.

While this rescue effort was in progress, another similar action was under way. Not long after its arrival on the MLR, Company E, 2/5 had spotted in the No-Man's-Land between the two main defensive lines a figure that appeared to be the body of a Marine. Since one man had been reported missing from an earlier 1st Marines patrol, recovery of the body, which had been propped up against a mound of dirt in the open, was undertaken. A special Company E patrol left the main line shortly before dawn on the 22d and reached the recovery area at daybreak. After artillery had laid down smoke, the patrol moved in, quickly recovered the body, and set out for friendly territory. Before the Marines had advanced very far on their return trip, the Chinese interdicted their route with heavy mortar fire, which killed one member of the patrol and wounded another. When the 5th Marines patrol returned to

---

[68] 1stMarDiv ComdD, Jun 52, App. I, p. 8.
[69] *Ibid.*

JAMESTOWN shortly after 0700, it carried not only the body it had recovered but also that of the Marine who had been killed on the recovery mission.

By the end of June, major command changes had taken place within the 1st Marine Division as well as in several other UNC components. On 13 June, Brigadier General Robert O. Bare took over the second spot from Brigadier General Twining. Both ADCs were graduates of the Naval Academy and both were native mid-Westerners (General Bare—Iowa, General Twining—Wisconsin). Before joining the 1st Marine Division in Korea General Bare had served at Camp Pendleton, California where most recently he had been commanding general of the Training and Replacement Command. His World War II experience included participation in both European and Pacific campaigns. He was the Staff Officer, Plans, in the U.S. Naval Section for the Allied naval group that planned the amphibious assault at Normandy, France. Later he served in the Peleliu and Okinawa campaigns and, with the ending of hostilities, had participated in the surrender and repatriation of the Japanese in north China.

The outgoing ADC, General Twining, was being reassigned to the Office of the Commandant, HQMC. For his outstanding service as assistant division commander from March through May 1952, he received a Gold Star in lieu of his second Legion of Merit with Combat "V."

Other high-level changes in command that had also recently taken place had included the UNC commander himself, General Ridgway, who had been succeeded in mid-May by General Clark. Major General Glenn O. Barcus, USAF, had assumed command of Fifth Air Force, replacing Lieutenant General Everest on 30 May. On 4 June, Vice Admiral Robert P. Briscoe had been named the new Commander, Naval Forces Far East to succeed Vice Admiral C. Turner Joy who had held the position since August 1949. And in I Corps, Major General Paul W. Kendall, USA, took over as corps commander on 29 June from Lieutenant General O'Daniel.

The end of the second year of the Korean fighting and the beginning of the third was observed by the Chinese with an attack against the 2d Battalion, 5th Marines, manning JAMESTOWN positions to the left of the regimental sector. Commanded at that time by Lieutenant Colonel Thomas J. Cross, 2/5 was new on line, having relieved 2/1 during the night of 15–16 June.

Late in the afternoon of 24 June, the enemy began registering his mortars and artillery on MLR company positions of 2/5 and a portion of the rear area occupied by the battalion 81mm mortars. Chinese incoming, sometimes intense, sometimes sporadic, continued until shortly after 2130. By this time the CCF were moving down their trenches toward a key outpost, Yoke, known also as Hill 159, which was still occupied on daytime basis by the Marines and lay north of the Company F Sector (Captain Harold C. Fuson). Moments later, the 34 men temporarily outposting Yoke saw the Chinese and opened with small arms fire, but the Marine positions were quickly enveloped by the Chinese. The Americans occupying the forward slopes of Yoke suffered many casualties from the intense fires supporting the enemy rush.

While the initial attack was in progress, the Chinese were able to position and fire machine guns from behind the outpost and in trenches on the forward slopes. Communist mortars interdicted the Marine supply routes to make normal withdrawal and reinforcement measures difficult. The Marines moved into bunkers, called down preplanned fires, and continued the defense. Although the Chinese had overrun Yoke, they could not evict the Marines. At about 0300, the enemy withdrew. When the 2/5 troops followed to reoccupy the forward slopes of Yoke, the enemy renewed his attack and struck again. As before, the Marines took to bunkers and called in defensive artillery fires. These boxing fires fell around the outpost perimeter until first light when the attackers withdrew for the second time.

Four other outposts in the battalion area were involved in the anniversary attack, but the action around Yoke was by far the heaviest. It resulted in 10 Marines of 2/5 killed and 36 wounded. At Yoke alone, 9 were killed and 23 wounded. Enemy dead were 12 known and 50 estimated killed. Chinese wounded were estimated at 100. At one point during the attack on Yoke, the outpost commander reported that the enemy were wearing gas masks and using tear-gas grenades. Investigation revealed that the Chinese had carried and even worn the masks, but that they had employed white phosphorus grenades rather than tear gas. This was the first instance Marine division personnel had ever encountered of CCF soldiers carrying gas masks in an attack and it was "believed part of the enemy's hate campaign to impress their troops with the possible use by the UN Forces of CBR (Chemical, Biological and Radiological) warfare." [70]

This violent eruption of enemy activity on the night of 24 June was followed by a brief period of greatly reduced ground action. Late on the 29th, however, the battlefront lull was broken when the 1st KMC Regiment sent out a raiding party to capture Chinese soldiers and their weapons and equipment, to inflict casualties, and to destroy positions. Second Lieutenant Kwak Sang In had his reinforced platoon from the 3d Company, 1st Battalion, equipped with rifles, carbines, machine guns, flamethrowers, and explosives. Target for the attack was an enemy outpost four miles south of Panmunjom that overlooked the Sachon River.

The patrol followed the general pattern of previous raids. It made use of supporting elements positioned on high ground in front of the objective. In this action the patrol struck from the rear, using artillery fire for both the assault and the withdrawal. Another similarity existed in that the results were nearly the same—no prisoners taken but fewer casualties to the attackers. One difference from earlier operations was that this patrol employed flamethrowers and TNT for destroying bunkers and inflicting casualties. Both weapons were credited in the killing of 12 and the wounding of 6 Chinese, in destroying 1 mortar and 7 bunkers, and in burning 3 other bunkers and numerous automatic weapons and rifles. Because of the heavy weight of a loaded flamethrower and the small size of the Korean Marines carrying these weapons, the flamethrower operators were fairly well exhausted by the end of the patrol.

A Long Fourth of July [71]

The approach of the American Fourth of July holiday marking an earlier struggle for freedom was appropriately accompanied by ground action initiated by all of the mainland MLR regiments. In the KMC area, a 3 July raiding party struck at forward enemy posi-

[70] Selden, *Div Staff Rpt,* p. 16.

[71] Unless otherwise noted, the material in this section is derived from: *PacFlt EvalRpt* No. 5, Chap. 8; and 1stMarDiv, 5thMar, 7thMar, 11thMar, 1/5, 2/5, 1/7, 3/7, 1st TkBn ComdDs, Jul 52.

tions before dawn, killing nine Chinese. In the center regimental sector Colonel Thomas C. Moore's[72] 7th Marines were also engaged in an active sector defense. In the left battalion spot 3/7, which had replaced 2/7 on line, dispatched raids on each of the first three nights of the month. Its Company G patrol on the night of 2–3 July was to be involved in one of the most costly small unit actions in the western Korea tour of duty for the Marine division.

Operational plans called for the platoon night raid on the 2d to be followed by a dawn attack the next morning. In both actions, the prisoner-taking aspect of the mission was considered a primary one. The early part of the operation was uneventful. One platoon moved forward toward the objective, Hill 159 (Yoke), 1,200 yards beyond combat outpost (COP) White, to the regimental left, without making contact with the enemy. The platoon then established a base of fire on favorable terrain from which the attack by the second platoon could be supported.

The second platoon passed through the forward position of the first shortly before 0630 and moved out into enemy terrain. It advanced less than 300 yards before its progress was halted by a Chinese force of battalion strength occupying the objective, Hill 159. Heavy enemy rifle and machine gun fire, hand grenades, mortar and artillery deluged the advancing Marines. Many of them quickly became casualties, but the operation continued, due in part to the determination and initiative of the NCOs. One of these was Staff Sergeant William E. Shuck, Jr., in charge of a machine gun squad. When the leader of one of the rifle squads became a casualty, Sergeant Shuck assumed command of that squad in addition to his own. Although wounded, he organized the two units and led them against the objective. Nearing the summit of the hill, the sergeant was hit a second time. Still he refused evacuation, remaining well forward in the lines to direct his assault force.

It was not until he had received orders to break contact with the enemy that the sergeant pulled back from the attack. During the withdrawal he looked after the other Marine casualties, making certain that all dead and wounded had been evacuated from the zone of action. While directing the last of the evacuation, Sergeant Shuck

[72] Colonel Moore took over regimental command on 11 June. The former CO, Colonel Honsowetz, had been named Assistant Chief of Staff, G-3 of the 1st Marine Division.

was struck by a sniper's bullet and killed by this third hit.[73] He was one of four Marines killed in the engagement. Forty others were wounded. Although no Chinese were captured, Marines estimated the enemy suffered losses of 50 killed and an additional 150 wounded.

To the east of the 7th Marines, the 5th Marines in the right MLR sector ordered a company-size patrol, also on the night of 2–3 July. Company A, 1/5 was directed to attack successively three outposts in the vicinity of the village of Samichon along the river bearing the same name and two miles beyond the point where the MLR crossed the river. After the reinforced company had taken the first two objectives, which were unoccupied, it received orders from division to return to the battalion area. Despite the fact the patrol had ventured far beyond the Marine lines, it did not come into contact with any Chinese forces.

A 2/5 combat patrol leaving the MLR just after dawn was successful in inflicting casualties on the enemy, taking prisoners, and destroying enemy field fortifications. The patrol made good progress until a Marine inadvertently set off an enemy mine. This mishap gave away the patrol's location and prompted reprisal by the Chinese. A one-hour fire fight followed. Then the patrol called in smoke and returned under its cover to JAMESTOWN. Marine casualties were 1 killed and 11 wounded. The second 2/5 patrol that same date was a successful ambush completed 10 minutes before midnight. In the brief clash that developed, Marine ambushers killed 6 enemy and wounded 8 more. The Marine force suffered no casualties.

The ambush patrol returned 15 minutes after midnight on 4 July. Even at that early hour division artillerymen had already initiated an appropriate ceremony to mark the Fourth. On 2 July, I Corps had directed the massing of fires on 4 July on the most remunerative targets in each division area. All objectives in the corps sector were to be attacked simultaneously at specified times for a one-minute period by employing a firing technique known as time on target

[73] The leadership, bravery, and unselfish devotion to duty earned for Sergeant Shuck the Medal of Honor, an award made to 14 Marines during the fighting in West Korea. During the earlier part of the war, 28 Marines had received the Medal of Honor. Of these, 17 were awarded posthumously. Five Navy hospital corpsmen, all attached to the 1st Marine Division, also earned the MOH. These awards, with one exception, were for heroism under combat conditions during the 1952-1953 period of the Korean War.

(TOT).[74] Normal daily fires were also to be carried out. Designated as Operation FIRECRACKER, the shoot expended 3,202 rounds in the division sector. Light and medium battalions of the 11th Marines, plus its 4.5-inch Rocket Battery destroyed some enemy trenches, bunkers, mortar and artillery positions, and damaged others. The division reported that the special fires on 4 July had also resulted in 44 known CCF casualties, including 21 dead, and 12 more who were estimated to have been injured.

More casualties, however, resulted from the issuing of another I Corps directive, this one dealing with the conduct of raids to seize prisoners, obtain information about the enemy, and to destroy his positions, supplies, and equipment. Back in June, the EUSAK commander had first stressed to his corps commanders the increased importance of combat raids to obtain additional intelligence during this period of stabilized conflict.

Although General Selden had submitted two division plans, he strongly believed that smaller patrols could accomplish the objective with fewer casualties and loss of life.[75] In particular, the division commander pointed out to I Corps that adequate defense of the 35-mile-long Marine division front did not permit the withdrawal of a sizable force for patrol missions without endangering the security of the entire Corps sector. The attack order was issued, however, on 3 July for the first large-scale raid to be conducted prior to 7 July. The code name BUCKSHOT 2B was assigned for this particular raid. As soon as he received the date of execution for the proposed operation, the Marine division commander advised I Corps that designation of 7 July as the cut-off date for the raid precluded proper rehearsal of attack plans. The operation would also conflict with rotation to the States of 2,651 Marines, whose replacements would not be available until 11 July. Corps turned a deaf ear; division then ordered a battalion-size attack for the night of 6–7 July.

Before dusk on 6 July, Lieutenant Colonel Daughtry's reinforced

---

[74] In the TOT technique, participating units time their initial volleys to ensure that their shells arrive on the target at the same time.

[75] Among division commanders in the I Corps area, General Selden was not alone in his grave misgivings of this method of gaining information about the Chinese. Major General A. J. H. Cassels, 1st Commonwealth Division, shared with the Marine commander the belief that such operations were too costly for the intended purpose. *McGill comments* and Brigadier C. N. Barclay, *The First Commonwealth Division: The Story of British Commonwealth Land Forces in Korea, 1950–1953* (Aldershot, England: Gale and Polden Ltd., 1954), p. 127, hereafter Barclay, *Commonwealth*.

1st Battalion, 7th Marines moved into position—on the left, a tank-infantry force, A/1/7 (still under Captain Thompson), to create a diversion; in the center, the main assault force, Company C (Captain Robert A. Owens); and on the right, a reinforced platoon from Company B (Captain Lyle S. Whitmore, Jr.) to support the attack by fire from positions close to the objective, Yoke. Earlier, three reinforced squads from Captain Thompson's unit had occupied combat outposts in the area of operations to deny the use of key terrain to the enemy and to provide additional fire support in the attack. At 2200, Captain Owens' Company C crossed the line of departure and set its course for Yoke, three-quarters of a mile northeast. Five minutes later the Company B support unit moved out to occupy the intermediate objective, COP Green, one-half mile southeast of Yoke. As it took up positions on COP Green, Captain Whitmore's Company B platoon discovered that no Chinese were in its vicinity; in fact, the platoon was not to encounter any enemy forces during BUCKSHOT.

Even though Company B failed to engage any Chinese, the remainder of the battalion encountered more than its share. About 450 yards southwest of the objective the Company C attack force was hit by an enemy ambush, which cut off Captain Owens' lead element. Although the Chinese directed strong efforts at halting the Marine advance, they were unsuccessful in this attempt. The Marines pressed the attack and seized Yoke 20 minutes after midnight.

On the left, the diversionary attack unit, Company A supported by the five tanks of the 2d Platoon, Company D, 1st Tank Battalion, and by a section of flame tanks from the armored battalion headquarters, began its mission at 2355. In three-quarters of an hour, the tank-infantry unit reached its objective, the first high ground southwest of Yoke. Tanks turned their 90mm guns on known Chinese positions on the hill to the north. During the next hour, the big guns of the M–46 medium tanks sent 49 rounds into enemy emplacements. The Marine tanks ceased fire at 0113 when Captain Thompson was alerted to assist Company C. He left one rifle platoon with the tanks.

Over on the high ground to the north and east, the attack force was under heavy fire from Communist mortars and artillery and was also receiving a number of enemy small-unit probes. At 0200, Company A made contact with Company C. Captain Thompson found the main force somewhat disorganized as a result of the wounding of the company commander, Captain Owens, the loss of several key officers and NCOs, and the effects of the lead element of Company C being ambushed and cut off. After being briefed on the situation by Captain Owens and conducting a reconnaissance, Captain Thompson recommended to the battalion commander that the entire force be recalled before daylight. At 0310 the two companies at Yoke began to disengage, returning to the MLR by 0636 on the 7th, without further casualties.

The one platoon of Company A and seven tanks of the diversion unit were still in their forward positions on the left and had prepared to resume firing. At dawn the M–46s relaid their guns on targets that had become visible. Tank gunners destroyed two observation posts and three machine gun positions and damaged many feet of trenchlines. At one point in the firing, the tank platoon commander, Second Lieutenant Terry K. Donk, using a power scope, observed ". . . two officers in forest green uniforms without equipment. They were definitely giving orders to machine gunners and infantry."[76] These 2 were among the 19 counted casualties (10 wounded) that the tankers inflicted during BUCKSHOT.

With the return at 0645 of the tank-infantry diversion force, the special operation for obtaining prisoners and information ended. No Chinese had been captured and no data gleaned from Communist casualties, listed as the 19 reported by the tankers and an estimated 20 more wounded or killed. Marine casualties from the operation were out of proportion to the results achieved—12 dead, 85 wounded, and 5 missing. It had been a high price to pay for a venture of this type, particularly when the primary objectives went unaccomplished.

During the entire 4–7 July period, 22 Marines had lost their lives in combat operations. Division reported that 268 Marines had been wounded during the long Fourth of July. These figures were the highest since September 1951 when large scale attacks by UN forces had first been abolished in line with the new tactic of positional warfare that would be waged until the truce talks resulted in an armistice.

---

[76] 1st TkBn ComdD, Jul 52.

Changes in the Lineup[77]

Division casualties were considerably higher during the first week in July than they were for the rest of the month. Once the pace of combat slowed, following the initial flurry of activity, the front again settled down to the patrol, raid, and ambush routine that had marked the static period of the Korean fighting. In accordance with the orders previously issued by higher authority the division placed continued emphasis on gathering all information it could about the enemy, his dispositions, and tactics. To assist in this effort, General Selden in July removed his reconnaissance company from defense of its small sector of JAMESTOWN and directed the unit to conduct training for its primary mission, obtaining intelligence about the enemy. Its place on the MLR was assumed by the two amphibian tractor companies then on line.

Another change of lineup took place on 14 July. At this time a battalion from the 15th Regiment, U.S. 3d Infantry Division took over the role of the maneuver element in the Kimpo Provisional Regiment, then held by 1/1, thereby releasing that battalion to its parent unit. With this change, the 1st Marine Division had a full regiment in reserve for the first time since its arrival in western Korea. A later shift in units occurred on 26 July when the 7th and 1st Marines traded places and missions. At that time the MLR, from west to east, was manned by the KPR, 1st AmTrac Bn, KMC, 1st Marines, and 5th Marines.

Opposing them in mid-July were an estimated 27 infantry battalions, whose primary missions were to defend the sectors assigned. The division credited these units with the capability of launching limited objective attacks at any time or of taking part in a major attack with a force of up to 57 infantry and 16 artillery battalions, augmented by 40 tanks or self-propelled guns. It was estimated also that the enemy could cross the Han in battalion strength in the vicinity of the northern shore of Kimpo Peninsula at any time and that Communist aircraft could attack anywhere in the division sector. Enemy forces identified at the end of July, from west to east, were the 193d, 195th, and 194th Divisions of the 65th CCF Army; the 189th Division of the 63d CCF Army; and the 118th Division, 40th

---

[77] The material in this section is derived from the 1stMarDiv ComdD, Jul 52.

CCF Army, which had recently moved from a position opposite the Commonwealth and U.S. 3d Infantry Divisions. Infantry strength of the Communists was established at 28,328.

Replacement and Rotation[78]

Marine infantry strength at the end of July 1952 was little more than half of the Chinese total. The division personnel strength was maintained by the monthly replacement and rotation program of Marines to fill vacancies created by the return of Marine combat personnel to CONUS (Continental United States) and combat losses. In the second quarter of 1952, the division rotated 433 officers and 6,280 enlisted men from Korea. In exchange, 506 officers and 7,359 enlisted men arrived from the States in replacement drafts. A new arrival could expect to stay with the division about 10½ months.

In the late spring of 1952 many of the division's new replacements were "dental cripples"—Marines requiring dental treatment, even emergency care in some cases.[79] General Selden directed that contact teams be formed to meet the replacement drafts in Japan. During the last leg of the trip to Korea dental personnel screened the new combat Marines on shipboard. By the time the division area had been reached, the dentists knew what remedial work would be required by incoming troops. At the end of the summer the problem was well under control.

Even though the 1st Marine Division in July continued to be somewhat in excess of its authorized strength in total personnel, it had certain imbalances and was in rather short supply of certain ranks and specialists. While the normal tour for most infantry officers ranged from 9 to 12 months, an excess of company grade officers, particularly lieutenants, had resulted in a reduction of the Korean tour for them to just six months. This brief period of duty plus an intra-division rotation policy that caused a mass shifting of duty assignments every three-to-five months tended to reduce unit combat efficiency. On the other hand the change of assignments had a favorable effect in that it broadened the experience of indi-

---

[78] Unless otherwise noted, the material in this section is derived from *PacFlt EvalRpts* No. 4, Chaps. 9, 10; No. 5, Chaps. 8, 9.

[79] *Brunelli ltr.*

vidual Marines. Beginning in the summer of 1952, however, the division modified this policy to reduce its number of intra-division transfers.

Personnel shortages existed in both the artillery and tank MOSs (Military Occupational Speciality). Mass rotation of reservist company grade artillery officers had necessitated the transfer of infantry officers to the 11th Marines for training and reassignment within the regiment. During the time when the supply of artillery officers was limited, however, the quality of support rendered remained high.[80] The other major shortage in the division was that of qualified crewmen — both drivers and gunners — for the M–46 tanks. Neither tank driving nor gunnery for the M–46 was taught in the tank crewmen's course conducted at Camp Pendleton, California. General Selden requested of Lieutenant General Franklin A. Hart (CG, FMFPac) that "tank crewmen be thoroughly trained prior to leaving the U.S." [81]

Fundamental to the tank problem was a shortage of the M–46 itself. At the training facility, Training and Replacement Command, Camp Pendleton, M–46 engines had been available for maintenance instruction but no tanks for the training of gunners and drivers.[82] General Hart pointed out this deficiency to the Commandant, General Lemuel C. Shepherd, Jr. On 13 August the Commandant directed the transfer of five tanks to the training installation from the 7th Tank Battalion,[83] also located at Camp Pendleton. At the same time General Shepherd ordered an increase in the school quota for tank crewmen. The first graduates would not reach the division in Korea, however, until the November draft.

The presence of not fully trained personnel in a combat zone was not limited to the division. In the summer and fall of 1952, a large number of volunteer reservists, both pilots and enlisted replacements with little experience since the end of World War II, joined the 1st MAW. It had not been possible for the Stateside training and tactical squadrons, themselves short of personnel and aircraft, to qualify all pilots as combat ready. It fell upon the wing in Korea, therefore, to take the needed corrective action. The more experienced

---

[80] *PacFlt EvalRpt* No. 4, p. 9–27.
[81] 1stMarDiv ComdD, July 52, p. 4.
[82] FMFPac ComdD, Jul 52, App VIII, Encl (7), Anx (E).
[83] FMFPac ComdD, Aug 52, App I, Encl (35).

1st MAW pilots, after completing their combat missions, flew instructional flights to help prepare the rusty fliers. Some reserve pilots, away from regular daily flying since 1945, found the adjustment too difficult and turned in their wings. MACG–2 operated "Pohang U," a training course for forward air controllers. In practically every squadron, there were shortages of electronics personnel. Jet squadrons found mechanics hard to come by. There were never enough motor transport replacements. For unqualified enlisted Marines, squadrons operated on-the-job training programs.

To maintain a reasonable degree of unit proficiency, the wing limited the monthly turnover of pilots to 25 percent. Like the division, the wing employed split tours between an officer's primary duty and staff work to broaden his experience. In some cases the amount of time required by administrative work as compared to a pilot's actual flying time reduced his proficiency in the air. In June, Task Force 95 reported that the proportionately large number of take-off and landing accidents on the carrier *Bataan* was caused by the rapid turnover of pilots and their need for frequent carrier qualification.[84]

A Marine pilot joining the wing could expect his assignment to last for 6 to 9 months. Personnel in a nonflight status had longer tours of 10 months to a year. Wing replacements were made on an individual basis, although there were plans that by mid-1953 a new policy of at least partial squadron replacement would be in effect. That 1st MAW squadrons were able to operate effectively on an individual replacement system was attributable to the peculiarity of combat conditions in Korea. Absence of real enemy aerial opposition permitted the use of basic, parade-type flight formations and nontactical approaches and attacks. An unusually high-level of experienced pilots in each of the two wing groups helped in the establishment of training programs and operational doctrine. The FAF limitation of four aircraft per flight eliminated the problem of large-scale, precombat squadron training as well as the difficulty of controlling and coordinating a large number of planes in a strike.

---

[84] *PacFlt EvalRpt* No. 4, p. 10–198.

Logistical Operations, Summer 1952[85]

Logistical support of the division and wing remained largely unchanged through July. Several modifications did take place, however, and these were:

(1) The change of responsibility for logistical support of ground-based units in Korea from Commanding General, 2d Logistical Command to the Commanding General, Korean Communication Zone (CG, KComZ).

(2) The opening of a pipeline system for resupply of aviation fuel at K-3, beginning in May.

(3) The beginning of increased support for airbase maintenance at those airfields housing Marine squadrons.

Resupply of common items used by both Marine and Army units was still being hampered by the Marines' limited knowledge of the Army supply system in effect and by their inability to obtain the catalogues, orders, and directives essential for requisitioning.

Two logistical operations, both of an engineering nature, took place between May and July 1952 in western Korea. One was Operation TIMBER, undertaken to provide lumber required to complete the bunker construction on the JAMESTOWN, WYOMING, and KANSAS lines. The division had estimated that three million linear feet of 4 x 8-inch timbers would be needed. Since lumber in this amount was not available through supply channels or standing timber in the division sector, Corps assigned the Marines a wooded area 50 miles to the east in the U.S. 45th Infantry Division sector. On 12 May a reinforced engineer platoon, under Second Lieutenant Roger E. Galliher, a truck platoon, and 500 Korean Service Corps (KSC) laborers,[86] began the cutting, processing, and hauling of timbers which were then trucked to the railhead. Between 500 and 1,000 logs were cut daily. When the operation ended in July a total of

[85] Unless otherwise noted, the material in this section is derived from: Selden, *Div Staff Rpt; PacFlt EvalRpts* No. 4, Chap. 9, No. 5, Chap. 8; 1stMarDiv, 1st EngrBn ComdDs, Jun–Jul 52.

[86] The KSC was a ROK quasi-military organization for logistical support of the UNC. Personnel were drafted from those rejected for Army service. Each KSC unit had a cadre of ROK officers and enlisted. All types of labor except personal services were performed by these Koreans. During its period in western Korea, the 1st Marine Division was supported by the 103d KSC Regiment of 5,222 men. CG, 1stMarDiv, *Civ Afrs and KSC*, pp. 8–9.

35,194 sections of timber had been cut. This was still not enough lumber to complete the required construction. Eighth Army then made up the difference, mostly with 12 x 12-inch timbers 30 feet long; these the Marine engineers cut to 4 x 8s for standard bunker construction.[87]

Operation AMAZON, published by I Corps on 12 June, ordered that bridging preparations be made for the approaching summer flood season. The previous August at the Honker Bridge, the one nearest the railhead, the Imjin had crested some 27 feet above normal. One reason for the precautionary efforts taken to insure bridge security during the flood season was the potential damage the Chinese could cause. Since they controlled the upriver area of the Imjin, before it entered the division sector, they could introduce floatable debris or explosives into the swift running flood waters. Another major concern was the logistical problem that would be faced by forward MLR units in event the bridges became impassable and the enormous strain that would thus be placed on helicopter resupply operations.

The I Corps directive specified that its divisions maintain a transport capability that would enable medium tanks to pass safely over bridges spanning the major rivers in their I Corps sector. The order also called for the removal of debris that could cause damage to bridges. Removal of those bridges vulnerable to flood conditions and the erection of emergency river spans were also to take place on corps order.

To carry out the I Corps operational order, General Selden put the division's own AMAZON plan into effect on 1 July. On this date Companies A, B, and D of Lieutenant Colonel Harry D. Clarke's 1st Engineer Battalion began extensive preparations for debris removal from the four bridge sites in the division sector. Even before this, Marine engineers and shore party personnel had been trained at special schools to handle U.S. Army equipment provided for the AMAZON operation.[88]

Beginning 1 June, division engineers began blasting away at objects

---

[87] Col Harry D. Clarke ltr to Hd, HistBr, G-3 Div, HQMC, dtd 1 Sep 66.

[88] This included employment of the 60-inch searchlight for night illumination, maintenance of boats for debris removal, and operation of the M-4 ferry. Other preparations by the division, of a non-engineer nature, included positioning of 13,000 life-saving floatation devices for use by frontline troops should they become shut off from planned evacuation.

that flood waters could loosen and carry into the bridge supports. Bridge approaches were improved and their supports strengthened. Each company had a detail living at the bridge site for which it was responsible. With the advent of heavy rains, these Marines were to operate 24-hour boat patrols to keep the river free of debris. The engineers were also to maintain a round-the-clock debris watch at the four division bridges—Freedom Gate, or the Munsan-ni Railroad Bridge in the left regimental sector; Honker and X-Ray in the center; and finally, Widgeon, very close to the Commonwealth boundary.

Heavy rains began on 27 July and continued until the 30th. On the first day the decking of Widgeon Bridge was completely submerged and Honker was removed to prevent its being carried away. Precipitation increased on 28 July and reached its peak on 29 July when 3.66 inches of rainfall were recorded. By the 30th, the rains had subsided but not before the overflowing Imjin had collapsed the X-Ray bridge. During the height of the four flood days, engineers fought the rains, flooding waters, and floating debris. The major effort took place downstream to save the Freedom Gate Bridge.

Assigned personnel removed debris from the bridge supports, guided large, dangerous pieces away with poles, while upriver the boat teams blasted still larger sections into manageable chunks that would pass between the bridge supports. These engineer efforts, in addition to regular repair and maintenance of the large road net, constituted the major ground activity in the 1st Marine Division sector in late July. August would bring more rains and emergency demands on the engineers, but the critical ground activity at that time would be directed against the Communists in the area around Bunker Hill.

CHAPTER III

The Battle of Bunker Hill

The Participants and the Battlefield—Preliminary Action on Siberia—The Attack on Bunker Hill—Consolidating the Defense of Bunker Hill—Company B Returns to Bunker Hill—Supporting Arms at Bunker Hill—In Retrospect

The Participants and the Battlefield[1]

THE TORRENTIAL RAINS that had fallen just before the end of July continued to affect ground operations into early August. Contacts between opposing forces were few and brief, and casualties remained correspondingly low. On 1 August, General Selden assigned the reserve regiment, the 7th Marines, the task of developing the secondary defense line, KANSAS, at the extreme right of the division sector. The 5th Marines, manning this regimental area and originally responsible for the construction, had been unable to reach the second line because bridging across the Imjin to the rear of the sector was washed out. By 3 July the division put a ferry service into operation at the site of the inoperable Honker Bridge for the purpose of feeding ammunition to combat units north of the Imjin. The critical resupply problem began to ease the next day when the waters overflowing the Widgeon Bridge further upstream receded sufficiently to permit restoration of normal vehicular crossings there.

Traffic in the air had, quite naturally, been less affected by the heavy rains and by the flooded, mucky terrain that was slowing ground movement throughout the entire division area. Flight operations during the first week of August produced a daily sortie rate

[1] Unless otherwise noted, the material in this section is derived from: *PacFlt EvalRpt* No. 5, Chap. 8; 1stMarDiv ComdDs, Jul–Aug 52; 1stMar, 2/1, 3/1 ComdDs, Aug 52; 1st MAW ComdD, Aug 52.

that would approximate the monthly average. In fact, the month of August was to become the record one for 1st MAW attack and fighter pilots during 1952, with a total of 5,869 sorties flown.

While the air people in August were maintaining a good weather pace against the enemy following the July downpours, the Communist ground troops apparently found the going too difficult to mount any sustained attack. The enemy merely continued his active defense, with an average of two contacts daily, while busily engaged in advancing his OPLR by creeping tactics. Even the usually assiduous Chinese artillery was strangely quiet. With respect to the enemy's excellent artillery capability, the 1st Marine Division in July learned that the Chinese had introduced a 132mm Russian rocket in their combat operations. The presence of this truck-mounted launcher, the *Katusha,* which could fire 16 rockets simultaneously, was indicated by a POW who had been informed by "his platoon leader that there were two *Katusha* regiments in the CCF." [2] In addition to this new enemy weapon, the Marine division reported the same month that positive sightings had been made of self-propelled guns emplaced well forward, and that there was an "indication that these guns were being used to fire direct fire missions from frontline revetments." [3]

Communist forward positions were gradually encroaching on JAMESTOWN. Since April 1952 the division had noted every month that the enemy was continuing to extend his trenches in the direction of the Marine MLR. The Chinese technique was to occupy key, high terrain at night, prepare the ground during darkness by digging trenches and constructing bunkers, and then vacate the area before daybreak. After nightly repetitions of this process had produced a tenable position, the enemy moved in and occupied it. By means of these creeping tactics, the Chinese hoped to acquire the dominating terrain necessary for controlling access to Seoul. The ultimate goal of the Communist forces was believed to be the 750-foot-high Paekhak Hill,[4] the Marine high ground position also known as Hill 229, just over a mile east of the road leading to Panmunjom and Kaesong.

---

[2] 1stMarDiv ComdD, Jul 52, p. 2.

[3] *Ibid.,* p. 1.

[4] CG, I Corps msg to CG, 1stMarDiv, dtd 18 Jun 52, in 1stMarDiv ComdD, Jun 52, App. I, p. 5.

During the four months that the 1st Marine Division's mission had been to conduct an aggressive defense of the EUSAK left flank, Marines had become familiar with a number of Chinese small unit infantry tactics. Shortly after assignment of the division to western Korea, General Twining, the ADC, had observed that the Chinese first made a diversionary frontal assault while the main force maneuvered around UNC defenders to attack from the rear. Almost invariably the Chinese employed this envelopment technique. Occasionally the enemy also used more passive measures, such as attempting to demoralize Marines in the front lines and subvert their allegiance by English language propaganda broadcasts. These attempts represented wasted effort. Not one Marine was swayed.

In some cases the Chinese were imaginative in changing their tactics or improvising new ones. This tendency had been noted as early as May by a 5th Marines battalion commander, Lieutenant Colonel Nihart, after 1/5 had engaged the enemy in a limited objective attack:

> . . . when friendlies marked targets with WP [white phosphorus], the enemy would immediately drop rounds of WP between the target and friendly troops to conceal the target and to confuse friendly FOs [artillery forward observers]; the enemy tried very hard to take prisoners (rather than shoot a friendly, they would often attempt to knock him out with a concussion type grenade); counterattacks were made in waves of four to seven men deployed in a formation somewhat similar to the Marine Corps wedge; snipers were deployed in holes that were mutually supporting; concerted efforts were made to knock out automatic weapons; . . . for close-in fighting, the enemy used PPSH [Soviet-made 7.62mm submachine gun] guns and grenades rather than bayonets; the enemy attacked behind well coordinated mortar fire; some enemy snipers were observed to have bushes tied to their backs. . . .[5]

On occasion 1st Division Marines found evidence that the enemy had infiltrated their lines. It appeared the most likely spot for line-crossers to make their way into the Marine rear area was from the far bank of the Imjin between the Sachon and Han Rivers where the enemy MLR was only a short distance from the sector held by the 1st Amphibian Tractor Battalion. Two enemy agents "armed with pistols of German manufacture, six hand grenades, and one set of field glasses"[6] had been apprehended here by a Marine reconnais-

[5] 1/5 ComdD, May 52, p. 12.
[6] HqBn, 1stMarDiv ComdD, May 52, p. 27.

sance company patrol. The prisoners had stated they were "part of a force of one thousand men who were infiltrating to form a guerrilla force somewhere in South Korea." [7] Six days later, after a brief fire fight between a small group of Chinese and a Marine outpost in the center of the division sector, the defenders discovered that two of the three enemy dead wore under their own clothing various articles of Marine uniforms. Neither of the Chinese had identification or any papers whatsoever. It was believed that both were enemy agents and that the attack on the outpost was a diversion "for the express purpose of detracting attention from infiltrators." [8]

Even though enemy tactics and attempts to penetrate Marine positions demonstrated a good deal of soldierly skills, his conduct of defensive operations was nothing short of masterful. This was especially true of Chinese construction of underground earthworks. It appeared that the Chinese had no single pattern for this type of field fortification. Like the Japanese in World War II, the Chinese Communists were experts in organizing the ground thoroughly and in utilizing a seemingly inexhaustible supply of manpower to hollow out tunnels, air-raid shelters, living quarters, storage spaces, and mess halls. Americans described the Chinese as industrious diggers,[9] who excavated quickly and deeply for protection against UN bombardments. From numerous reports of ground clashes in the 1st Marine Division sector and from observations made by Marine pilots, it became known that the enemy was quick to seek cover whenever he was exposed to sustained artillery bombardment or air attack.

What was not known, however, was the extent of these subterranean shelters. One Chinese account, allegedly written by a reconnaissance staff officer named Li Yo-Yang, described the protection of a CCF shelter to a recently captured UN prisoner as they were under Allied artillery bombardment. While shells exploded all around the position the enemy boasted: "There's no danger of being killed on a position fortified by the Chinese People's Volunteers . . . Don't

[7] *Ibid.*

[8] 1stMarDiv ComdD, Jun 52, p. 5.

[9] "The Chinese attack by 'shovel' proved effective and difficult to combat. They burrowed forward almost continuously, even under direct observation. Every foot of advance provided added opportunity to attack Marine COPs with greater impunity. While this activity possibly provided Marines with target practice in both small arms and mortars, these CCF working parties in a narrow trench 7 to 10-feet deep probably took very few casualties." Col William R. Watson, Jr. ltr to Hd, HistBr, G-3 Div, HQMC, dtd 18 July 67.

you know it's impossible for your shells to penetrate our air-raid shelters?"[10] An American report on enemy field fortifications estimated that the amount of earth cover in Chinese air-raid shelters was as high as 20 feet, and in frontline defensive positions, up to 33 feet.[11]

Marine defensive installations carved out of the ground were not so extensive as those of the enemy opposing JAMESTOWN. "In spite of orders, instructions, and inspections many bunkers were only half dug in, then built up above the ground with sandbags," observed one Marine battalion commander.[12] Back in April, just after the Marine division had settled in the west, its 1st Engineer Battalion, using U.S. Army drawings, had published bunker construction plans. Express instructions to frontline units were to "construct bunkers to provide simultaneously living and fighting space. Overhead cover on all bunkers will be such as to withstand direct hit from 105mm and to allow friendly VT fire over position."[13]

Some officers felt it was, perhaps, the work-during-light, patrol-at-night routine that resulted in the shallow draft Marine bunkers. Others suggested that the relatively limited defensive training received by the more offensive-minded Marines created a natural apathy to digging elaborate fighting positions.

It took a hole 12 feet square and 7 feet deep to house the Army, Lincoln-logs-type bunker the Marines first used in the spring of 1952. The fortification, using tree trunks up to eight inches in diameter, had a cover of seven to eight feet. This consisted of four feet of logs, and three-to-four more feet of rocks, sandbags, and earth fill. By the summer of 1952, the division developed its own style of bunker, a prefabricated timber structure designed to fit into a hole eight feet square and somewhat less than seven feet deep. This size fortification could accommodate a .50 caliber machine gun, crew members, or several riflemen. Provision was also made for the inclusion of a sleeping shelf in the rear of the bunker. Its construction required no saws, hammers, or nails, only shovels to excavate. The

---

[10] *A Volunteer Soldier's Day: Recollections by Men of the Chinese People's Volunteers in the War to Resist U.S. Aggression and Aid Korea* (Peking: Foreign Languages Press, 1961), p. 193, hereafter CPV, *Recollections*.

[11] *PacFlt EvalRpt* No. 5, p. 8–90.

[12] LtCol Roy J. Batterton, Jr., "Random Notes on Korea," *Marine Corps Gazette*, v. 39, no. 11 (Nov 55), p. 29, hereafter Batterton, *Korea Notes*.

[13] CO 5thMar msg to 5thMar units, dtd 20 Apr 52, in 5thMar ComdD, Apr 52, #2, App. II, p. 6.

major drawback to erection of the prefab was the difficulty in manhandling the heavy roofing timbers, 11 feet long, 12 inches wide, and 4 inches thick. On top of this was placed a two-foot layer of sandbags, tarpaper covering, and a four feet high layer of earth that completed the structure and partly camouflaged it.

Battlefield construction was carried out by the infantry regiments to the limit of unit capabilities. The division engineers, one company per frontline regiment, augmented at times by shore party units, supplied the technical know-how and engineering materials and equipment. These combat support troops processed the lumber for bunker construction and built fortifications for forward medical treatment and one bunker for observation of battle action by civilian and military dignitaries, irreverently called VIPs (Very Important Persons), who frequently visited the division. Engineers also erected some of the barbed wire barriers in the forward areas and, when necessary, cleared firing lanes for weapons housed in bunkers.

The processing of timbers for easier and faster bunker-construction had begun on 28 July, but this was hardly in time for the most difficult fighting the division had faced thus far in western Korea. Given the name Bunker Hill,[14] this battle would take place in the center sector of the division line manned since 27 July by Colonel Walter F. Layer's 1st Marines.[15] On that date Lieutenant Colonel Armitage's battalion, 3/1, took over from the 3d Battalion, 7th Marines on the left, and 2/1 (Lieutenant Colonel Roy J. Batterton, Jr.) relieved the 2d Battalion of the 7th Marines on the right.[16]

Across No-Man's-Land, units of two Chinese divisions faced the 3,603 men of the 1st Marines. From west to east opposite the Marine regiment's frontline battalions were elements of the 580th Regiment, 194th Division, 65th CCF Army and of both the 352d and 354th Regiments, 118th Division, 40th CCF Army. The 352d Regiment held most of the area on which the battle would be fought.[17] Enemy combat efficiency was rated as excellent and his forward units were well-supplied. The Chinese conducted an active defense, using limited

[14] Since bunkers were in everyone's mind and frontline units were heavily involved in the bunker-construction program, it is felt likely "someone in G–2 arbitrarily assigned the name." Col Gerald T. Armitage ltr to Hd, HistBr, G–3 Div, HQMC, dtd 6 July 67, hereafter *Armitage ltr*.

[15] Two days earlier Colonel Layer had taken over the command from Colonel Flournoy.

[16] Lieutenant Colonels Gerald F. Russell and Anthony Caputo, respectively, commanded 3/7 and 2/7 at this time.

[17] 1stMarDiv PIR 657, dtd 13 Aug 52.

objective attacks, numerous small-size probes, and creeping tactics to extend their OPLR line. Communist soldiers offered well-coordinated and tenacious resistance to Marine patrols, raids, and attacks. Within enemy lines a 775-foot elevation, known as Taedok-san, was situated directly north of the Marine division center and commanded the entire Bunker Hill area.

On JAMESTOWN, the dominating height was Hill 201, 660 feet high[18] and immediately to the rear of the MLR in the left battalion sector. Southwest of this elevation was the Marine stronghold, Hill 229, just 23 feet lower than Taedok, and believed by the Marines to have been the objective of the August battle. Directly north of Hill 201 was Hill 122, adjacent to the enemy OPLR, and called Bunker Hill by the Marines. It was shortly to become the scene of bitter fighting. The crest of Hill 122 was about 350 yards long. At a distance of about 700 yards, it generally paralleled the northeast-southwest direction of JAMESTOWN in the left of the 2/1 sector and adjoining 3/1 sector.

Southwest of Bunker and a little more than 200 yards from the Marine MLR was Hill 124. This Hill 124–122 axis, for tactical purposes, was known as the Bunker Ridge. The ridgeline, roughly "cashew" in shape almost anchored back into the MLR on the forward slopes of Hill 229. To the northeast of Bunker Hill and separated from it by a wide saddle[19] was another enemy position, Hill 120. (See Map 9, for outposts and key hill positions in the 1st Marines center regimental area in early August.)

Approximately one mile east of Hill 124 was Hill 56A, or Samoa, the right flank limit of the immediate battlefield. It guarded the best avenue of approach into the Bunker Hill area, the Changdan Road. Another Marine position west of Samoa was Hill 58A, or Siberia, a sentinel overlooking a long draw running down the east sides of Hills 122 and 120. Both Samoa and Siberia were outposted by squads. Another 1st Marines squad occupied Hill 52, on the other side of Changdan Road and not quite a half-mile east of Samoa. The entire battlefield was cut up by numerous gullies and draws, most of which paralleled Bunker Hill.

---

[18] Frequently cartographers use elevations for names of hills. Heights on the Korean maps are in meters, and many of these hills derive their name (i.e., number) from their elevation. For changing meters to feet, the conversion factor 3.28 is used.

[19] A saddle, the low point in the crest line of a ridge, is much in appearance like the side view of a riding saddle.

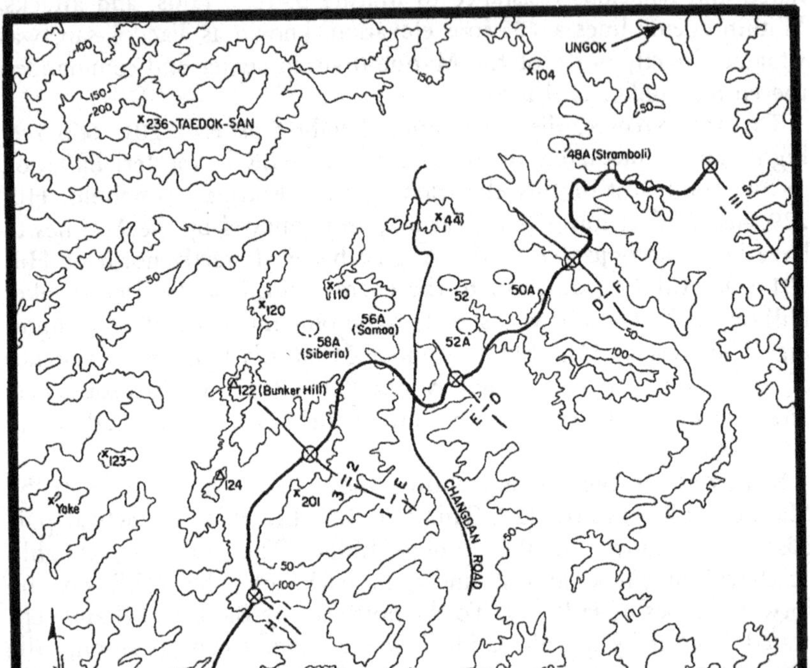
IST MARINES SECTOR
OF JAMESTOWN
(Division Center)
8 AUGUST 1952
Outposts in right battalion sector
Hill designation
MLR
0
500
1000
1500
Yards
UNGOK
104
236 TAEDOK-SAN
48A (Stromboli)
44
110
120
52
50A
56A
(Samoa)
58A
(Siberia)
52A
122 (Bunker Hill)
123
124
Yoke
201
CHANGDAN ROAD
229
KMC
100
150
200
50

MAP 9

K. White

Preliminary Action on Siberia[20]

The first round in the battle of Bunker Hill began as the fight for Siberia, Hill 58A. Just slightly more than a quarter of a mile from JAMESTOWN, this squad-size outpost, the most western in the right battalion sector, had been occupied in June when the division moved its MLR forward. Since Siberia was located halfway between the Marine MLR and the Communist OPLR, the Marine seizure of Siberia prevented the Chinese from holding terrain suitable for employing 60mm mortars against Marine frontline troops.[21] Strong enemy outposts on Hills 120 to the north and 110 to the northeast constantly threatened the squad on 58A. From these two forward positions, Chinese troops early on 9 August 1952 streamed down to Siberia, launching in the process the Bunker Hill battle.

Just before 0100 an estimated four enemy squads fell upon Hill 58A, outposted by Company E Marines. Using assorted infantry weapons, the raiding party forced the outnumbered Siberia occupants to withdraw. By 0145 the outpost Marines returned to the MLR. At this time the JAMESTOWN sector south of the outpost, also held by Captain Jesse F. Thorpe's Company E, was under attack by approxmately 50 Chinese.

After breaking up the enemy assault by well placed friendly mortar fire, the Marines enjoyed a brief respite from Chinese pressure and formulated plans to recapture Siberia. It was decided that a reinforced Company E platoon would counterattack to regain the outpost. At 0355, the 11th Marines fired a five-minute preparation against the objective. On schedule, the platoon crossed JAMESTOWN at 0400 and in the darkness headed towards the outpost. Advancing carefully to avoid detection as long as possible, the Marines reached the area near the base of the hill by 0525. Heavy enemy artillery and mortar fire again forced the Marines to withdraw, and the platoon returned to its company CP at 0545. So far, the 58A action had resulted in the wounding of 32 Marines and the killing of another.

It became evident that more preparation, by Marine air and artil-

---

[20] Unless otherwise noted, the material for this section is derived from: 1stMarDiv ComdD, Aug 52; 1stMarDiv G-3 Jnls, 9–11 Aug 52; 1stMar, 1/1, 2/1, 3/1 ComdDs, Aug 52.

[21] 1stMarDiv ComdD, Aug 52, App. VII, p. 1.

lery, would be required for the recapture of Siberia. At 0650, four Marine F9F jet fighters worked the hill over with napalm and 500-pound bombs. Three hours later, a flight of Air Force F-80 "Shooting Star" jets dropped eight 1,000-pound bombs on the same target. With the aerial attack complete, Marine artillery opened fire. Five minutes later another Marine reinforced platoon launched a second ground attack. This was made by a unit from Company A (Captain Robert W. Judson) of the regimental reserve battalion, supported by a Company E platoon. Again the Marines advanced to the open sector south of the hill before the enemy reacted. As before, the Chinese response was a devastating barrage from their supporting weapons. The stubborn Marine assault against Siberia brought down the full weight of Chinese firepower—rifle, machine gun, and hand grenades—but the attack force would not be beaten off. At 1103 the Siberia hill again belonged to the Marines. Quickly the Company A platoon began to organize a defense to repulse the Chinese counterattack, which was certain to come.

In anticipation of a prompt and violent retaliation by the Chinese, and to help the speedily improvised defense efforts, the 2/1 battalion commander, Lieutenant Colonel Batterton, had sent forward the supporting platoon from Company E. This reinforcing unit reached Siberia within seven minutes after the Marine attackers had gained possession of the objective. The new arrivals scarcely had time to dig in before a hail of mortar and artillery shells forced all the Marines to seek cover in a defiladed position on the southern side of the slope. From here, the 2/1 force directed counter mortar and artillery fire onto the top and far side of Siberia and unleashed their own assault weapons against the Chinese soldiers pressing for possession of Siberia. By midafternoon, with heavy enemy counterfire on the position and their casualties reaching nearly 75 percent, the Marines were forced to withdraw and return to their own lines. The hill had changed hands twice and the enemy had employed 5,000 rounds of artillery in the contested ownership.

Badly mauled by two actions against Hill 58A, Company E came off the lines to reorganize, exchanging positions with Company A, of Lieutenant Colonel Louis N. King's 1st Battalion. About this time Company C, less one platoon, had moved from the 1/1 rear area forward to an assembly point behind 2/1 in preparation for a night counterattack to retake the now battle-scarred outpost. Without the customary artillery preparation, the attacking force at 2245 crossed the MLR at a point directly south of the former outpost Samoa, which had been abandoned earlier when Siberia fell. Working their way northwest towards Siberia, the Company C Marines, commanded by Captain Casimir C. Ksycewski, cautiously approached the assault line. Reaching it at 0105 on 10 August the force deployed immediately and rushed the objective.

At about this time the Chinese defenders opened fire but could not halt the assaulting Marines. The struggle to regain the Siberia objective was fierce; some of the Chinese refused to yield and fought to their death. Most, however, held their defense positions only briefly before retiring to the refuge offered by the reverse slope of the hill. Gaining the crest of Hill 58A at 0116, the Company C commander ordered a platoon to the other side of the objective to dispatch remaining elements of the enemy force. The resulting fire fight lasted nearly four hours. At daybreak, however, the enemy, in estimated company strength, strenuously renewed his counterfire and, for a third time, forced the 1st Marines to retire from the disputed hill and return to the main line.

Later that day, at the regimental CP, Colonel Layer called a staff conference to decide on the best course of action. Successive Marine withdrawals had been caused by the intense enemy shelling. The key to its effectiveness was the observation provided the Chinese from Hills 122 and 110. Heavy enemy fire had also caused most of the casualties, 17 killed and 243 wounded, in 1st Marine ranks. It was decided to shift the battle area to better restrict this enemy capability not only to observe Marine troop movements but also to call down accurate fire on friendly attacking units. Bunker Hill, an enemy outpost west of Siberia, was selected. In the eyes of 1st Marines tacticians, possession of Hill 122 instead of Hill 58A presented three major advantages:

Hill 122 offered excellent observation into the rear of enemy outposts;

Possession of Hill 122 would greatly strengthen the MLR in the regimental sector, effectively neutralize Siberia, provide dominating terrain that was more defensible than 58A; and

Bunker offered an excellent opportunity for an attack employing the element of surprise against the enemy.

To help preserve this tactical surprise, the plan for the Bunker Hill attack included a diversionary attack against Siberia. Making this secondary effort would be a reinforced rifle platoon and a composite unit of gun and flame tanks. For the main attack, Lieutenant Colonel Batterton's 2d Battalion would employ a reinforced rifle company with supporting artillery and armor, if needed. The operation was to be conducted at night, to further ensure the opportunity for tactical surprise. For the same reason, the attack was not to be preceded by artillery preparation on either objective. To the right of the 1st Marines, however, Colonel Culhane's 5th Marines would support the diversion by artillery and tank fire placed on enemy strongpoints in the Ungok area, about 1¼ miles northeast of Siberia. During daylight, air, artillery, and tanks attacked targets on both 122 and 58A. Priority of effort in the 1st Marines area went to units preparing for the Siberia-Bunker offensive.

The Attack on Bunker Hill[22]

At dusk on 11 August, 1,000 yards behind the MLR in the western sector of the 2/1 line, the eight Company C tanks that were to provide much of the diversionary effort at Hill 58A moved out of their assembly area. Leading the column east of the MSR, Changdan Road, were four M–46 mediums, mounting 90mm guns. They were followed by an equal number of flame vehicles. Each M–46 was specially equipped with an 18-inch fighting light, actually a searchlight with a shutter over the lens, to be used for battlefield illumination. The flame vehicles, World War II M4A3E8 mediums, mounted a 105mm howitzer in addition to the flame tube. As the tanks reached the Changdan Road, they turned north, crossed the MLR, and proceeded to preselected positions. (See Map 10.)

When the M–46 gun tanks were in position to fire on Siberia and its flanks, their powerful 90s opened up on the objective. At this time, 2110, the first section of flames (two tanks) made its way

[22] Unless otherwise noted, the material for this section is derived from: Encl (1) to CG, FMFPac ltr 0762/161 over A9 to CMC, dtd 25 Nov 52, Subj: "Summary of 1stMarDiv Sit from 20 July–20 Oct 52," hereafter FMFPac, *1stMarDiv Sum, Jul–Oct 52;* 1stMarDiv, 1stMar, 2/1, 1st TkBn ComdDs, Aug 52.

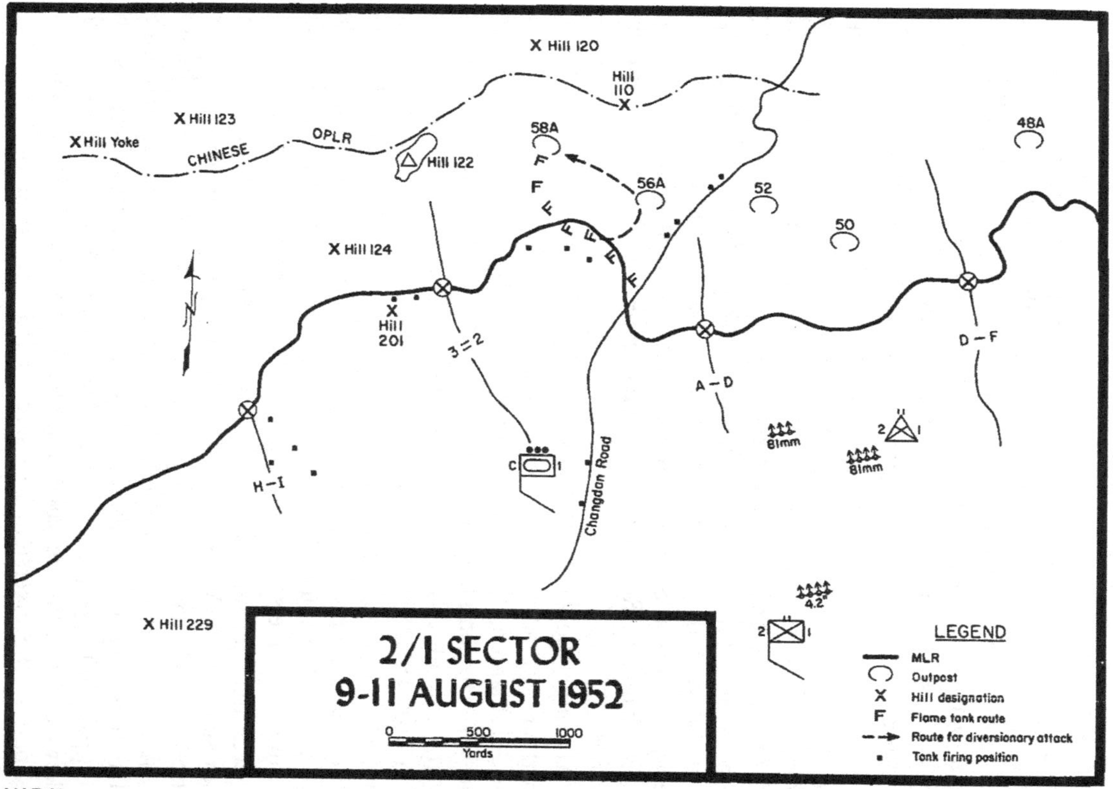

MAP 10

K WHITE

along the stream bed between the MLR and Hill 56A (Samoa). Lighting their way with very short bursts of flame, the two tanks advanced in this manner to the base of Hill 58A. There the vehicles paused momentarily, then began to move up the near slope, using longer spurts of flame to sear the ground and sparse vegetation to the crest of the position. The gun tanks, in the meantime, had shifted their fire from Siberia northeast to neutralize Hill 110. When the flame vehicles reached the top of Siberia, they lumbered down the far slope, firing then in shorter bursts and sweeping the area with machine guns to discourage any enemy infantry interference.

With some fuel reserved to light their way on the return trip, the flame section reversed its course from the far side of the objective, mounted the crest, and clanked back to the Changdan Road. When the first section had returned, the second departed, completing its mission in much the same manner. Tank personnel of both groups observed that the enemy artillery and mortar fire was medium to heavy on Siberia. Some rifle fire was also received. Gun tanks, firing from Changdan Road east of Siberia, experienced less fire from the Chinese.

Although the flame vehicles had completed their mission and were on their way home, the M–46s remained on position in support of the 3d Platoon, Company D which, at 2230, was advancing from the MLR to complete the infantry part of the diversion. Staying out of the low ground that the tanks had used, the platoon swept over Hill 56A at 2255 and immediately struck out for the further objective, Hill 58A. Gun tanks firing their 90s on the Chinese OPLR on Hill 110 and on Siberia illuminated the target area with their fighting lights, the shutter of which the tankers flicked open and closed during each five-second interval that the light remained on.

Less than an hour after crossing JAMESTOWN, the platoon from Captain George W. Campbell's Company D reported the capture of Siberia. The enemy quickly made his presence felt at the objective; a half hour before midnight, he assaulted the hill in reinforced platoon strength. Ten minutes later the Chinese withdrew and the Company D Marines, in accordance with their battle plan, did likewise. At about the same time the 5th Marines, having completed its part in the diversion, also secured from the operation.

Ten minutes after the diversionary infantry had cleared Samoa while enroute to Siberia, the main attack force, Company B, which had come under operational control of 2/1 at 1800, crossed the MLR, the line of departure. Moving at a fast pace to preserve the element of anticipated surprise, the attack force, commanded by Captain Sereno S. Scranton, Jr., soon deployed two squads of the lead platoon against the near side of the hill. By 2318 on 11 August the squads were moving up Bunker Hill and, 10 minutes later, one platoon had gained the top of the objective and one was at the base of the hill, both moving northward along the forward slope. As the advancing units neared the end of their sweep forward, they began to come under small amounts of rifle fire from the front and left flank of the position.[23] The Company B platoons continued to advance, returning well-placed small arms fire.

Soon the intensity of Chinese small arms fire increased; at the same time enemy mortars and artillery opened up on the company. Marines attempting to assault the top of Hill 122 also came under a hail of hand grenades hurled by the staunch Chinese defenders. After a brief but vicious fight at point-blank range, the Chinese gave ground on the eastern side, heading uphill. Several Marines pursued the fleeing enemy to the summit, then joined the rest of the assault units of Company B in organizing a defense. By 0300, 12 August the battle had quieted down and for a short while all firing ceased. Then, as the Marines began to dig in, a bypassed pocket of enemy resistance came to life. Two fire teams in the 1st Platoon took these Chinese Communists under fire.

Even as the fighting continued, Marines and KSC personnel were hauling fortification materials towards Bunker to consolidate the precarious foothold. For a while, enemy mortars unleashed a heavy fire against the newly won position, but by 0230 Company B was able to report that enemy shelling had stopped and that the objective was in friendly hands. A new fire fight broke out at 0345 between a small force of enemy soldiers occupying a draw forward of Bunker Hill and Marines nearby. The exchange of fire continued for nearly two hours, but short of harassing the Marines on Bunker Hill the enemy did not launch a counterattack. Dawn on 12 August

---

[23] Recalling the Marine seizure of Bunker, the G–3, 1stMarDiv at that time expressed the view that "taking these places was easy but holding them under heavy Chinese artillery and mortar fire was extremely costly. Our counterbattery fire was ineffective because we were limited to from one to eight rounds per tube per day, depending on the weapon, by Army order, because of an ammunition shortage." Col Russell E. Honsowetz MS comments, dtd 15 Jun 67, hereafter *Honsowetz ltr II.*

revealed that thus far in the Bunker Hill fighting 1 Marine of Company B had been killed and 22 were wounded. The earlier diversionary attack on Siberia had resulted in only one Marine casualty, the wounding of the platoon commander, Second Lieutenant James W. Dion.

Personnel losses were kept to a minimum by the well-organized medical support and the efficient service of medical and evacuation personnel. A forward aid station was established in the vicinity of the Company E CP. Casualties that were not ambulatory arrived at this two-bunker installation usually by hand litter, other wounded men were transported in armored personnel carriers, U.S. Army tracked vehicles similar in appearance to the Marine LVT, that had accompanied the diversionary unit and were part of the Panmunjom rescue force stationed in the area of COP 2 on the 3/1 left flank. Wounded Marines were examined immediately. Minor injury cases were treated and discharged; more seriously injured personnel were given emergency treatment and evacuated. Movement to the rear was accomplished by ambulance jeeps. Helicopters, landing only 30 yards from the station, flew out the critically wounded. A sandbag-protected squad tent was used to house casualties waiting to be examined. This emergency aid station closed down on 13 August, when action in the right battalion sector diminished.

Even though the remainder of the morning of 12 August was practically free of any retaliatory attempts by the Chinese against Bunker Hill, the Marines occupying the new position were not idle, for they anticipated an immediate and severe reaction for capturing the hill. Quickly, but methodically, the company dug in. At noon, regiment passed to 3/1[24] the responsibility for Bunker Hill and operational control of Company B. Consolidation of Hill 122 continued until about 1500, when the Marines were forced to put down their entrenching tools and grab their rifles instead. The Chinese had suddenly launched an intense mortar and artillery attack against the hill. Defending Marines expected to see enemy soldiers start up the western slopes at any minute.

Actually, more than an hour elapsed before the Communists initiated their first main ground attack to regain Bunker. By that

---

[24] Initially the diversionary attack against Siberia and subsequent assault against Bunker had been made by Marines of 2/1 since Siberia was in the 2/1 sector. On 12 August operational control was transferred to 3/1 as the fighting continued at Bunker, in the area of responsibility of the left battalion sector.

time, heavy casualties from the continued shelling had forced Company B to pull back from the ridge and take up positions on the reverse (eastern) slope of Bunker Hill. At this point, with reduced Company B forces and with no radio communication between Captain Scranton's unit and 3/1, Lieutenant Colonel Armitage sent I/3/1,[25] under Captain Howard J. Connolly, forward from the MLR. Shortly before 1600, a force of more than 350 Chinese lunged out of the low ground of Hill 123, west of Bunker, to attack defensive positions along the ridge between Hills 124–122. Striking in rapid succession first the west side and then the northern end of the Company B position to find a weak spot in the defense, the enemy counterattack finally concentrated on the southwestern part of the hill.

An intense exchange of fire raged here until 1715, when the defending fire of Company B plus the added weight of the Company I reinforcements combined to stall the enemy advance. Having failed to gain their objective, the Communists abruptly broke off their artillery and mortar fire and ordered their infantry to withdraw. They pulled back only to the far side of the hill, however. By 1740 the enemy was occupying his new post on the northern slope, while the Marines continued to hold their positions on the reverse slope of Bunker Ridge. Enemy supporting fires had lifted and a lull ensued in the fighting.

Consolidating the Defense of Bunker Hill[26]

Even before the Chinese had made their coordinated attack against Hill 122 in the midafternoon of the 12th, the 1st Marines had implemented a plan of action to assure that the critical position would remain in Marine hands. In addition to the movement of Company I/3/1 to reinforce Bunker Hill and of Company I/3/7 as its relief on the MLR, a precautionary displacement was also made of the 3/1 reconnaissance platoon to Hill 124 to tie in that terrain

---

[25] From the division reserve, Captain Anthony J. Skotnicki's company, I/3/7, was en route to take over the I/3/1 sector. As an interim measure, Captain Byron J. Melancon's Company H extended its MLR positions to the right to cover the Company I area.

[26] Unless otherwise noted, the material for this section is derived from: 1stMarDiv ComdD, Aug 52; 1stMarDiv G-3 Jnl, 12–13 Aug 52; 1stMar, 1/1, 3/1 ComdDs, Aug 52.

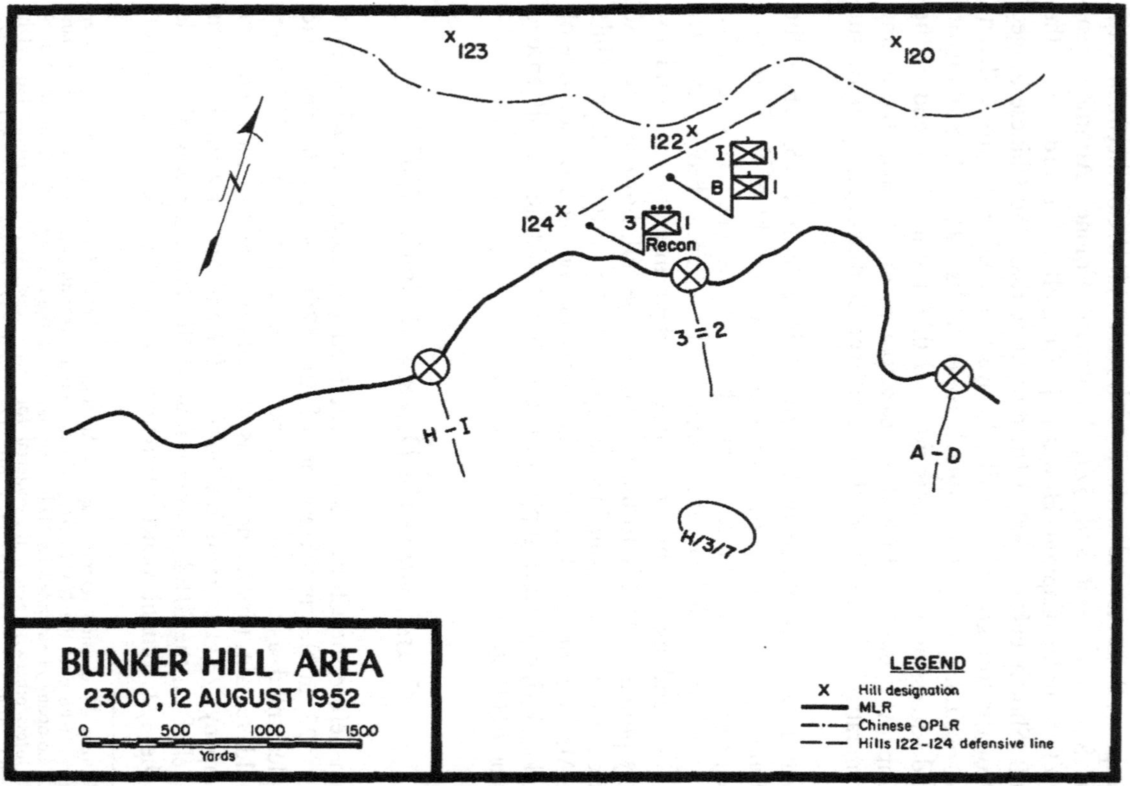

MAP 11

K WHITE

feature with both Bunker and JAMESTOWN and thus consolidate the defense north of the MLR and west of Bunker. (See Map 11.)

Other activities behind the line aimed at making the Marine position on the newly seized hill more tenable. As one step in this direction, General Selden shifted most of his reserve into the zone of action. Before the end of the day remaining units of 3/7 were placed under operational control of 3/1, and 2/7 was attached to Colonel Layer's reserve. The 7th Marines was directed to place its 4.2-inch mortars on call to the 1st Marines. Priority of artillery support went to the Bunker Hill regiment. Within the 1st Marines, the regimental commander moved two provisional platoons (118 Marines) of the reserve 1st Battalion to the 3d Battalion sector. All 81mm mortars in 1/1 were sent to the left battalion. The fire plan also called for employment of all the 60mm mortars that could bear on the crest of 124–122, with 81mm and artillery box-me-in barrage fires on the ridge and flanks.

Machine guns from the MLR were assigned missions on the crest of Bunker Ridge and 4.5-inch ripples were planned on the deep enemy approaches. Gun and flame tanks were to protect the right flank of Hill 122 where the steep draw between Bunker and the MLR offered the most dangerous approach into Bunker Hill. Supplies and fortification materials, meanwhile, were being carried forward and casualties taken to the rear by the relief party. Although 3/1 initially reported that the Bunker Hill fighting had resulted in 58 killed or wounded Marines, a later battalion count showed this number to be 34—5 killed and 29 wounded.

Most of the casualties had been caused by hostile shelling. Although the Hill 122 reverse slope afforded some cover from the Chinese artillery and mortars, the positions on the crest did not offer any protection, so Marines continued their trenchworks and other defensive preparations at a rapid pace and supporting fires were registered by 1900. The approach of night was certain to bring renewed Communist attempts to capture Bunker Hill.

At 2000, Lieutenant Colonel Armitage reported to division that his force on Hill 122 occupied the entire reverse slope and that the Marine of I/3/1 and B/1/1 were digging in and consolidating their scant defenses. Enemy shells were still falling on both Bunker and Hill 124. Company commanders forward of the MLR estimated that as many as 400 Chinese occupied the ridge on the other side of the slope from the Marines. Since the crest of the long Hill 124–122 ridgeline was fairly level, the gentle incline of the Bunker rear slope permitted defending Marine units excellent fields of fire to the ridge crest, a major consideration in the 3/1 battalion commander's decision to adopt a rear slope defense. Moreover, the top of the ridge could be swept with direct fire from the MLR as well as supporting weapons from the two nearest companies on JAMESTOWN. Opposing Marine and Chinese forces were thus lined up for a continuation of the battle for Bunker Hill.

It appeared that the Chinese wished to attempt a diversionary tactic of their own. To draw attention away from Hill 122 they engaged a Marine outpost east of Bunker and a KMC ambush far to the left before attacking Bunker again. In the KMC sector, shortly after 2300, an enemy infantry platoon walked into a trap near the eastern edge of the Sachon and 500 yards south of the Munsan-ni-Kaesong rail line. The brief fire fight lasted only 10 minutes before the Chinese broke contact.

Perhaps the ambush was incidental to the forthcoming attack against the Bunker complex, but this same reasoning cannot be applied to the Communist-inspired action which broke out shortly at Hill 48A, Stromboli, another friendly outpost far to the east of Hill 122. Near the right limiting point of Colonel Layer's 1st Marines and the 5th Marines boundary, Stromboli was another Marine fire-team-by-day, squad-by-night position. It occupied a small rise 250 yards forward of the MLR and commanded the immediate sector in all directions. The entire MLR in the regimental right was dominated by the enemy-held Hill 104, a half-mile north of 48A.

Communist infantry opened the attack without benefit of any supporting arms preparation and rushed to seize Hill 48A early on the morning of 13 August, a few minutes after midnight. Defending Marines immediately responded with small arms and automatic weapons fire. By the time the outpost commander had informed battalion of the attack by radio, the far right sector of the 1st Marines line, held by Captain Clarence G. Moody, Jr.'s Company F, had also come under attack. Firing rifles and submachine guns and hurling hand grenades as they assaulted the main position, the Chinese attempted to penetrate the JAMESTOWN defenses. In spite of the enemy's concerted efforts, the Marine line remained staunch.

At Stromboli, the Communists met with no greater success,

although they did cause enough casualties to warrant the dispatch of a Company F reinforcing squad. When this unit left the MLR, at 0106 on 13 August, the Marine line was still under a heavy attack not only from Chinese infantry but from hostile artillery and mortars as well. Out at Hill 48A the outpost remained in comparative quiet until the approach of the reinforcing party. As the Company F squad neared the base of the hill, Chinese infantry that earlier had been assaulting the Marine MLR turned their rifle and machine gun fire from positions on the JAMESTOWN side of the outpost. A heavy rain of devastating mortar fire engulfed the reinforcing Marines. On order, they broke off the approach march and returned to the company rear area.

On the main line, meanwhile, Company F positions were still being bombarded by Chinese artillery and assaulted by their infantry. Casualties along the entire line forced Lieutenant Colonel Batterton to order his 1st Provisional Platoon, Headquarters and Service Company, 2/1, to the Company F command post. After the clutch unit departed the battalion area, at 0210, and approached Captain Moody's CP, enemy fires immediately intensified. A violent fight erupted to the left of the Company F sector, but the Marines there held. The Chinese then tried to punch holes in other parts of the company line, moving eastward along JAMESTOWN. Each failure to breach the line seemed to signal a decrease in the intensity of Chinese shelling.

This easing of Communist pressure against the main line enabled the Company F commander to put into operation a new attempt at the reinforcement and rescue of Stromboli. After the initial enemy assault in the early hours of 13 August had ended in failure, the Chinese made repeated attempts to capture the outpost. At one time it appeared that a company of Chinese had overrun the hill. Later, however, the Stromboli stronghold radioed that the enemy force, subsequently identified as only a platoon, had encircled the Marine position. To relieve enemy pressure at Hill 48A, Captain Moody employed a rifle platoon which set out for the outpost at 0325.

Simultaneously, as if their intelligence had advance knowledge of the 1st Marines recovery plan, the Chinese stepped up the tempo of their attack at Stromboli. A fresh assault by the enemy was stymied by Marine superiority in hand-to-hand combat. Thereafter, close-in defensive fires continued to ring the outpost and to discourage future assaults. The approach of the second Marine rescue party eliminated much of the pressure that Communist foot soldiers had maintained around the hill position. After a 90-minute exchange of fire with the enemy, the friendly platoon penetrated the encirclement and rushed to the besieged outpost at the hill crest. At this point the Chinese disengaged and withdrew towards the north.

After their diversion against Stromboli had approached the proportions of a full-scale attack, with the enemy having reinforced from platoon to company size, the Chinese then initiated their main thrust, an attempt to retake Bunker. Captain Connolly (I/3/1) had reported that shortly before 0100 Communist mortar fire had begun falling on his positions on the southern slope of Bunker Hill. Simultaneously, Chinese artillery stepped up the rate of its barrage fires as did the assaulting close-in enemy infantry. Captain Connolly then requested the 11th Marines to place box-me-in fires around the Marine company positions on Hill 122. Artillery furnished these defensive fires almost immediately.

Shortly after 0130, the Marines in the center and right of the I/3/1 position observed a large number of Chinese, deployed into a skirmish line, headed directly for their part of the hill. The attack was accompanied by intense machine gun and rifle fire. It was countered by an equally heavy reply from Marines on Bunker. For nearly four hours the battle raged at Hill 122. Unsuccessful enemy frontal assaults were followed by attempts to dislodge the defenders from the rear. In their continuing thrust against the hill, the Chinese were repulsed by Marine coordinated support fires—tank, rocket, artillery, and mortar.

By firing on known or suspected assembly areas and Chinese infantry units advancing up the draws towards Hill 122, these Marine supporting weapons helped to preserve the status quo at Bunker. Repeated box-me-ins were also fired by the 11th Marines during the early-morning Communist attacks on 13 August. Exploding friendly mortar shells increased the effectiveness of the hill defense; nine rocket ripples[27] fired by the artillery regiment further supported Marines at the critical terrain position. Tanks unleashed

---

[27] A characteristic of 4.5-inch rocket launcher is the discharge of 24 rounds in quick succession, called a ripple. A battery of six launchers can fire 144 rounds on target in less than a minute.

their deadly fire on nearby enemy outposts to neutralize them; their 90mm guns, aided by the battlefield illumination from tank fighting lights, helped eliminate Chinese foot soldiers attempting to envelop Marine positions on Bunker.

It was in this direction that an enemy force, estimated at reinforced battalion strength, headed during the early morning fighting on Hill 122. At 0330, the struggle for possession of the height had reached the climax. For an hour the issue remained in doubt. Then, as the Chinese small arms fire decreased, the tempo of the enemy's artillery shelling increased. This, the division correctly deduced, announced the beginning of a temporary Communist withdrawal from Bunker Hill.

Although the immediate danger of the enemy onslaught had ended for the time being, Marines to the rear of the JAMESTOWN Line stepped up their defensive preparations. Division, regimental, and battalion operational plans were put into effect to prevent a Chinese victory. The seriousness of the situation on the 1st Marines right flank at Stromboli early on 13 August had resulted in the movement of one company of 5th Marines into blocking positions behind the MLR near the left regimental boundary. To the south of Bunker Hill, relief and replacement units from the division reserve, ordered into action late the previous day, maneuvered into position to strengthen the regimental front. One of these relief units, G/3/7, under command of Captain William M. Vanzuyen, had just deployed from its assembly area to pass through the ranks of an MLR company and take over the Bunker Hill positions. The Marines' situation on Hill 122 had deteriorated so rapidly, however, that the 3/1 commander rushed two reinforced squads forward from I/3/7, the nearest MLR unit.

The Company G reinforcement unit jumped off from JAMESTOWN and arrived at Bunker shortly after sunup, where it reinforced Captain Connolly's positions during the height of the battle for possession of Hill 122. Not long after, the Chinese initiated their withdrawal under cover of increased artillery and mortar barrages. As they left, the Communists policed the battlefield in their typically thorough manner. A Marine platoon that swept the northern slope of Bunker failed to find any enemy bodies in this area so recently abandoned by the Chinese, but did take under fire and kill seven enemy that had remained on Hill 122.

Before I/3/1 had sent one of its platoons to reconnoiter the far side of Bunker Hill, Lieutenant Colonel Armitage ordered H/3/7, under Captain John G. Demas, forward to relieve friendly forces at the contested height. The exchange of units was completed before noon of the 13th. By late afternoon, except for Company H, all 2d and 3d Battalions, 7th Marines units that had moved up to reinforce the 1st Marines were on their way back to the regimental reserve area. At this time the 1st Marines CO, Colonel Layer, reported to General Selden that the Bunker Hill action during 12–13 August had resulted in 24 Marines killed and 214 wounded. On the right, in the 2d Battalion sector, an additional 40 Marines were listed as casualties, including 7 killed in the Stromboli defense. Chinese known dead numbered 210, plus an estimated 470 killed and 625 wounded.[28] Artillery and aerial observers reported that between 1500 on the 12th and 0600 the following morning an estimated 5,000 to 10,000 rounds of enemy fire had fallen on 1st Marines positions, the "heaviest incoming fires received by the Division since coming into the present sector."[29]

The number of casualties from the Bunker Hill action was to increase further that same day with a renewed attack on the outpost. Before the Chinese again engaged Hill 122, however, they made a diversionary attack on the western flank at the extreme left of the 3/1 sector. At dusk on 13 August, the enemy shelled the Company G Marines at COP 2, the critical height overlooking the Panmunjom peace corridor. The shelling caused several casualties and lasted 90 minutes. Towards the end, Communist infantrymen moved forward and fired on the outpost. At about the same time, Company H personnel emplaced on the MLR to the rear of COP 2 began to receive artillery rounds in preassault proportions.

A ground attack in this western end of the 3/1 sector did not materialize, however. Instead, the Chinese resumed their attack on Bunker Hill. Since their temporary withdrawal early on the 13th, the CCF had repeatedly sent mortar and artillery barrages against the bastion to harass its new occupants. On occasion these well-aimed mortar rounds found their mark. Mortars interdicting a trail used for resupply of the Hill 122 defenders did inflict some casual-

---

[28] 1stMarDiv PIR 658, dtd 14 Aug 52.
[29] Selden, *Div Staff Rpt*, p. 19.

ties on two groups rushing emergency supplies forward from the MLR.

At 2100, while continuing his shelling of the left end of the 3/1 sector, the enemy lifted his preparation on Hill 122 to permit a CCF reinforced company to make a new assault there on the Marine defenders. Captain Demas called for box-me-ins to seal off his positions and illumination shells to help locate the enemy force. Utilizing the draw to the east of Hill 58A, the Chinese proceeded west to Bunker where they pitted one platoon against the center of the Company H, 3/7 line and another against the right flank. Defensive fires momentarily held off the intruders, although some were able to break through to the Marines' fighting positions.

Those enemy troops who penetrated the Marine defenses were quickly eliminated by grenades and small arms fire. Unable to weaken the Marine defenses any further and by now sustaining sizable casualties from unrelenting Marine artillery and mortar concentrations, the Communists withdrew at 2215. Marine defenders estimated they had killed 175 enemy during this latest encounter; a firm count of 20 bodies were found on the shell-torn slopes. Company H casualties, all from enemy mortar and artillery fire, were 7 killed and 21 wounded.[30] Enemy incoming was again heavy during this period, with a reported 3,000 rounds falling in the sector.

In the 3d Battalion sector, Marine and KSC stretcher bearers brought casualties to the I/3/1 CP, several hundred yards to the rear of the front line. At the command post, the critically wounded were airlifted by helicopter to the rear. Less seriously wounded casualties were placed in jeep ambulances and carried to the battalion forward aid station, about two miles away. Here a team of doctors and corpsmen examined and treated patients, discharged a few, but prepared most for further evacuation. At the 1st Marines forward aid station, patients were reexamined and their wounds redressed when required; discharge or further evacuation was also accomplished. Most of the Marines brought to this forward facility

---

[30] During the fighting on the 13th, Hospitalman John E. Kilmer was mortally wounded while "administering aid to the wounded and expediting their evacuation." Though wounded by enemy mortars, he continued his life-saving efforts until another barrage took his life. He had died shielding a wounded Marine undergoing emergency treatment. Hospitalman Kilmer, a distant cousin of poet Joyce Kilmer, became the first of four corpsmen serving with the 1st Marine Division to be awarded the Medal of Honor during the trench warfare in western Korea.

had become exhausted from vigorous activity in the high temperature and humidity which characterized the South Korean summer. The regimental aid station treated these heat cases and then released them to their units.

Company B Returns to Bunker Hill[31]

Division intelligence subsequently reported that the 2100 attack on 13 August had been made by an enemy battalion with a reinforced company in assault. This same unit again sent a small band of Chinese soldiers against Hill 122 at 0225 the following morning. This clash was to be the briefest of all offensives for control of Bunker Hill during the 11–17 August period. Prior to launching this four-minute fire fight, an enemy machine gun at Siberia had attempted to harass the Marines at Bunker Hill. In retaliation, Marine tanks illuminated this enemy weapon with their searchlights and immediately took it under fire with their 90mm guns, knocking it out of action. At the same time, enemy artillery attempted to shell friendly tanks. During this brief fire exchange, one tanker was wounded slightly and the lens of one fighting light was splintered by fragments from enemy shells bursting around the tanks. The inconsequential probe was made, Marines believed, not so much to seriously challenge Marines holding Hill 122 as it was to retrieve CCF dead and wounded from the major attack a few hours earlier that night.

Anticipating that a much heavier ground attack was close at hand, the 1st Marines ordered a reinforcement of the Bunker Hill position. Even before the heavy action on the 13th, this machinery had been set in motion. To this end, the 3d Battalion was to reinforce the Bunker defense by sending a 1/1 platoon to the hill and the 2d Battalion was instructed to return Company A (minus this platoon) to the reserve battalion. At 0415 on the 14th, Company E/2/1, led since 10 August by Captain Stanley T. Moak, took over from A/1/1 the responsibility for the 2d Battalion's MLR "Siberia sector," adjacent to the Bunker Hill area held by the 3d Battalion. The Company A reinforcing platoon arrived at Hill 122 just before dusk,

---

[31] Unless otherwise noted, the material in this section is derived from: 1stMarDiv, 1stMar, 3/1, 1st TkBn ComdDs, Aug 52.

preceding another CCF company attack by only a few hours.

At midnight the 1st Marines front was suspiciously quiet for a few minutes. Forward on Hill 122, there was no apparent enemy activity. Captain Demas sent out a two fire-team patrol from Bunker to reconnoiter northwest of Bunker towards the Chinese lines. Shortly after the eight Marines returned with a negative report of contact with the enemy, the regiment received a report about the outbreak of a small arms clash between defenders on the left flank of Bunker and an enemy unit farther west. At 0118 on 15 August what had initially appeared to be a minor contest suddenly erupted into a heated fire fight all along the 124–122 Bunker Ridge complex. At the request of Captain Dumas, Marine artillery fired protective boxes around the Bunker positions. This defensive maneuver held the attackers in check.

At this moment, Chinese infantrymen in the draw running alongside the 124–122–120 ridge system were massed for an assault on Bunker from the northeast. The plan might have been successful had not a fighting light from a tank on the main line intercepted the Communists in this state of their preparations. In a matter of moments, friendly artillery, mortar, and tank fire struck the Chinese and scattered the formation.

After discovering he could not successfully pull a sneak attack, the enemy reverted to his usual procedure, employing a preassault bombardment prior to his infantry assault. This preparation began at 0206; it reached the rate of approximately 100 rounds of 82 and 122mm mortar shells per minute. While supporting weapons pounded the Marines, the Chinese assault commander reorganized his attack force that the Marine shelling had scattered. Communist infantry then moved forward and fired on the Bunker Marines, who replied with rifles and machine guns and box-me-in fires. Unable to penetrate this protective mask around the positions, the Chinese gradually decreased their small arms and artillery fire until, at 0315, the rate of exploding shells at Hill 122 had dropped to only four or five per minute. Soon thereafter the small arms fire slacked off entirely and by 0400 even the mortars had stopped. Across the entire 1st Marines front, all was quiet again.

During the Company H defense of the hill, enemy losses, caused mostly by friendly artillery and mortar fire, were placed at 350, including 40 counted dead. Captain Demas' Marines suffered 35 casualties, of whom 7 were killed. En route to the MLR after relief by B/3/1, the company suffered four more casualties, including two KIAs, all the victims of Chinese mortars.

It was not long before these weapons inflicted casualties on Company B, which had six of its men wounded even before the H/3/7 unit had reached JAMESTOWN. Another Marine at Bunker was wounded by enemy mortars later that morning. At 1640 the Communists again probed Bunker Hill, this time in company strength. Striking in daylight during a thunderstorm and without any preparatory fires, the Chinese attackers failed to achieve any tactical surprise. The defenders fired both infantry and supporting weapons; some threw grenades at the few Communists who did manage to get close to the fighting positions. At 1750, the Chinese withdrew, this time leaving 35 of their dead in the attack area. Four Marines had been wounded; five others suffering from battle fatigue were later evacuated.

Exactly when the enemy would strike next at Bunker Hill was not known by the Marines. Most believed that the Communists would return but only speculated as to when. Although the battalion felt that "the enemy was not expected to attack again for some time,"[32] events were to prove otherwise. In any case, the battalion was prepared, having an adequate force on Bunker and sufficient local reserves to absorb an attack up to the strength of any received so far. Division supporting arms were readily available for commitment at critical points.

The Chinese soon put an end to the conjecture about the next attack. At 0040, 16 August, an enemy force, later estimated as a battalion, came out of positions to the west and north of Hill 122. Supported by mortars at first, and later on by artillery, the battalion sent one company against the Marine outpost. Several attacking elements were able to penetrate the defensive fires. These Chinese reached the crest of the hill and began using their rifles, automatic weapons, and hand grenades against the defenders. Captain Scranton called for reinforcements. A platoon from I/3/7 was dispatched promptly from the 3/1 sector. The reinforcements departed JAMESTOWN just as the fire fight on Bunker began to subside. By 0315, the enemy had begun his withdrawal, and another reinforcing element,

---

[32] 3/1 ComdD, Aug 52, p. 4.

1/3/1, had moved forward, this time from regiment to Lieutenant Colonel Armitage's CP.

About two hours later a brief fire fight flared up in the Company B sector. No ground assault was made on Marine positions. The enemy force, of undertermined strength, never closed with the Marines and within 10 minutes, the firing stopped. No casualties to the Marines resulted during this exchange. The earlier clash had resulted in the death of 3 Marines and the wounding of 27. Enemy losses were estimated at 40 killed and 30 wounded.

Before it came off the hill, Company B was engaged by enemy fire three more times. At 1945, Chinese mortars (82mm) wounded two Marines. Later, heavier mortars placed 20 rounds on Hill 122, but these caused no casualties. There were some losses, however, early on the morning of the 17th when C/1/1 was relieving the Bunker defenders. Captain Scranton's Marines sustained five more wounded from automatic weapons, five during the relief.

The second relief of Company B on Bunker brought to a close the battle that had been waged for possession of the vital hill complex. During the Hill 122 tours of Company C and other 1st Marines units that followed in August, seven more ground actions tested the Bunker Hill defenses. Only one of them, during the night of 25–26 August, was of significant size. This attack also failed to dislodge the Marines from the hill.

Supporting Arms at Bunker Hill[33]

It was quite natural that the flurry of ground activity during the battle of Bunker Hill created a need for increased participation from Marine supporting arms. The magnitude of infantry action during the contest for Hill 122 resulted in a monthly record to date in 1952 for the amount of air support received as well as the volume of both artillery and tank fires supporting the division. During this critical 9–16 August period, the 11th Marines played a part in every ground action except the feint attack on Siberia and the seizure of Bunker Hill, both of which were purposely executed without an artillery preparation. Medium tanks fired day and night missions during

[33] Unless otherwise noted, the material in this section is derived from: *PacFlt EvalRpt* No. 5, Chaps. 8, 9; 1stMarDiv ComdD, Aug 52; 1stMarDiv G–3 Jnls, 4–16 Aug 52; 1st Mar, 1st TkBn ComdDs, Aug 52; MAGs–12, –33 ComdDs, Aug 52.

most of the infantry action. Close air support at times amounted to a strike every 20 minutes.

During the ground action around Bunker, Siberia, and Stromboli, the division received close air support in amounts unparalleled for JAMESTOWN Marines to that time. Marine and U.S. Air Force pilots flew a total of 458 missions (including 27 ground controlled MPQ–14 radar bombing attacks) during five of the most critical days, 9–13 August. On two of them, the 1st Marine Division received priority of close air support along the whole EUSAK front. Fifth Air Force assigned 1st MAW aircraft to Marine requests for close air support as long as Marine aircraft were available.

The initial air strike by Marines in the Bunker fighting was on 9 August in support of counterattack plans for Siberia. MAG–33 provided a morning and evening flight of four F9F jet fighters to destroy enemy forces and defensive works on 58A (Siberia). USAF fighter-bombers attacked Siberia and other outposts nearby and enemy artillery positions supporting the Chinese forward line. On the next day, air operations, concentrating on Siberia, were stepped up considerably against enemy outposts. Thirty-five aircraft in nine missions attacked 58A with bombs, rockets, and napalm. These strikes were carried out by MAG–12 and U.S. Air Force pilots at irregular intervals during daylight hours. Air controllers reported good results. Other aircraft hit known mortar locations capable of supporting the Chinese. During the morning, Marine Attack Squadron 121 (Lieutenant Colonel Philip "L" Crawford) bombed and burned Bunker Hill. Just before sunset, F–80 and –84 jets of the U.S. Air Force dropped 15 tons of bombs on mortar positions and troops on and around Hill 120. Four F–80s also participated with eight Marine AD–2 propeller-driven attack aircraft in the morning attack on Bunker.

Air activity in support of the 1st Marines continued unabated on 11 August. Before the diversionary ground attack just after dusk that day, Marine and Fifth Air Force fliers repeated the treatment that Hills 58A and 122 had received the previous day. During daylight, supporting weapons positions were hit by FAF fighter planes. At night, MAG–12 air attacks guided by the MPQ–14 radar bombing system destroyed hostile artillery and mortars. Also during the dark, the medium bombers of the FEAF Bomber Command struck deeper in the rear at heavy weapons locations.

These Air Force bombers conducted four more controlled-bombing attacks against Chinese artillery during the early hours of 12 August, when Company B was consolidating its positions and hastily organizing the defense of Bunker Hill. After daylight and until dusk, MAGs–12 and –33 and USAF squadrons provided four-plane flights to strike troop assembly areas, supporting weapons positions, and observation posts close to Hill 122. In late afternoon, Marine pilots in four F9F Panther jets and three ADs bombed and burned the enemy side of Bunker Hill during the shelling and subsequent ground attack against the Marines on the eastern slope.

Marines flew, on 13 August, all of the daylight close air support missions in support of the actions on both Bunker in the center and Stromboli in the right of the 1st Marines sector. On 13 August, a total of 94 aircraft were committed over the regimental sector to conduct strikes in support of ground operations. Enemy Hill 104, commanding the 2/1 outpost on 48A (Stromboli), received four attacks. Fighter bombers (F4U propeller-driven Corsairs) carrying napalm, rockets, and 1,000-pound bombs, raided the hill mass at 0535. The other strikes against this key terrain-feature were made by attack and fighter aircraft during the afternoon. Other targets on the regimental right were weapons positions beyond Hill 104 and an enemy outpost one thousand yards west of Stromboli.

Most of the air support received by the 1st Marines on the 13th was directed against targets that were participating—or that were capable of taking part—in the battle on Bunker Hill. Against the enemy on the height itself, the Marines directed only three strikes, and these came late in the morning. A majority of the air attacks were dispatched against observation and command posts and the firing positions of both automatic and large caliber weapons. Chinese artillery and mortar fire had inflicted more casualties and punishment on the Marines than the enemy infantry assaults. As a consequence, the main effort of the close air support strikes was directed against these hostile supporting weapons.

After dark on the 13th, VMF(N)–513 commanded by Colonel Peter D. Lambrecht,[34] took up the air offensive against the heavy

---

[34] Two days later, Colonel Lambrecht, flying a F3D twin jet night fighter with his radar operator, Second Lieutenant James M. Brown, disappeared while on a night flight. The last known position of the plane was over the Yellow Sea, 50 air miles west of Pyongyang. At about that point the aircraft faded from the radar screen. Efforts to re-establish communications failed. It was reported that observers at sea sighted a crash and explosion at about this same time. Extensive search failed to uncover any trace of the Marines or their aircraft.

firing positions in the rear of the enemy line. The squadron conducted four attacks with its night fighters. Two of its attacks were made just before sunrise.

During the remainder of the battle of Bunker Hill, the ground fighting subsided and the requirement for close air support abated accordingly. On the 14th, only four daylight strikes were flown in the 1st Marines area. These, all by Marine squadrons, were against active artillery and mortars in the defilade of Hill 120 and others to the west on the far slope of Hill 123, and Chinese outpost positions, west of 48A, which had been pestering the Stromboli garrison. There were no flights after dark on the 14th, but on the following night, two MPQ missions were flown by VMF(N)–513. Each was a single plane flight against a reported artillery location. This was the final night air action in the battle for Bunker Hill. Daylight missions in support of Hill 122 defense after the sharp decrease of attacks on the 14th numbered only seven attacks, each by four planes. These, flown by Marines, continued to emphasize the destruction of enemy artillery.

Marine artillery continued its support of ground troops and air strikes. Cannoneers of the 11th Marines fired 21 flak suppression missions during the five days beginning on 11 August. This type of close coordination between Marine supporting arms further reduced combat losses of aircraft providing CAS to the division. The Marine artillerymen had played a vital part in the defense of the besieged outposts. Lieutenant Colonel Armitage credited the box-me-in fires with an important role in thwarting each enemy attack on Bunker.[35]

In the 24-hour period beginning at 1800 on 12 August, Marine artillery directly supporting the 1st Marines fired 10,652 rounds. Most of the ammunition was expended in support of the Bunker Hill defense; some was used in behalf of the Marines outposting Stromboli during the Communists' early morning diversion that day. On the 9th, the direct support battalion, 3/11 (Lieutenant Colonel Charles O. Rogers), had fired about one-fourth of the 12–13 August total. Many of the shells that first day of the Bunker battle were preparatory to counterattacks for regaining Siberia.

When the retaking of Hill 58A was discarded in favor of the surprise attack on 122, the amount of artillery support was reduced,

---

[35] 3/1 ComdD, Aug 52, pp. 3–4.

during the 1st Marines infantry preparations on the 10th and 11th, in keeping with the fire support plan. Upon seizure of Bunker, Lieutenant Colonel Rogers' business immediately picked up and quickly reached a crescendo the following day, when the 10,652 shells fired became a Marine one-day battalion record for western Korea until the last stages of fighting in 1953. Other Marine artillery battalions fired reinforcing missions during the critical period as did the 4.5-inch Rocket Battery which fired a large number of on-call ripples. The regimental commander later recalled that "during some of the crises every gun that could bear on Bunker in the 11th Marines and reinforcing units was shooting there."[36]

After a sharp drop on the 14th, the artillery support gradually decreased in proportion to the amount and strength of the enemy's action against Hill 122. By 20 August, 3/11 was firing only 244 rounds a day. Only on the 26th, during a serious Chinese attempt to retake Bunker, did the number of artillery rounds match the intensity of the fire support rendered during the earlier part of the month.

It was not only the quantity of 11th Marines support that the infantry called for during the battle of Bunker Hill; quality was equally important. A majority of the more than 28,000 rounds that 3/11 fired during the eight days of Bunker Hill fell around the besieged outposts. Many rounds were fired in defense of MLR positions. In both of these types of protective fires, extreme accuracy and precision were required due to the proximity of enemy and friendly lines in order to prevent any "short" rounds from falling among Marine positions. Lieutenant Colonel Armitage recalled that during the height of the battle on the night of 12 August, "we did have a bad scare . . . when Captain Connolly reported that friendly mortar fire was falling short."[37] The battalion immediately ceased fire with its 60mms, 81mms, and 4.2s and each piece was checked; the culprit was quickly located and within 5–10 minutes 3/1 resumed fire.

During the August battle, artillery in general support of the entire division and I Corps artillery reinforcing the fires of

---

[36] BGen Frederick P. Henderson ltr to Hd, HistBr, G–3 Div, HQMC and MS comments, dtd 20 Jun 67, hereafter *Henderson ltr III.*

[37] *Armitage ltr* and comments, p. 12.

Colonel Henderson's regiment, stepped up their efforts to destroy the distant and more difficult targets, including mortars and artillery. These continued to be the main cause of Marine casualties. Some of the labors of the 11th Marines gun crews did silence enemy heavy weapons, but personnel losses from enemy shellings still mounted, especially in the infantry units. To assist in the location and destruction of the enemy artillery, aerial observers spent considerable time in spotting and fixing Chinese weapons positions.

Besides these counterbattery efforts, the 11th Marines employed other artillery means to provide the additional support the 1st Marine Division requested during Bunker Hill. Two of these were the counter-counterbattery and the countermortar programs, the former being a passive defense-deception program to minimize Chinese counterbattery fires against 11th Marines weapons. Nearly every day C Battery, 17th Field Artillery Battalion, fired special request missions.[38] Another type of fire, flak suppression, aided the cause of close air support pilots delivering ordnance against those Chinese positions taking Bunker Hill and Stromboli under fire. At night, illumination shells helped outpost and frontline Marines in locating groups of enemy massing for assault on Hill 122.

Mortars (4.2-inch) of the 1st Marines contributed heavily to the defense of the outposts. Operations reached a peak on 12–13 August when, in a 24-hour period, Captain Carl H. Benson's mortar company fired 5,952 rounds—4,084 high explosive and 1,868 illuminating. In addition to their defensive fires, these hard-hitting weapons attacked Chinese mortars, automatic weapons, defensive positions, and troop formations with deadly accuracy.

No less precise and lethal were the fires of Captain Gene M. McCain's gun tanks (Company C, 1st Tank Battalion), and the battalion flame tanks. Three of the latter had fired their 105s in support of the KMC on the morning of the 9th before the vehicles received orders to move east to join Company C temporarily. On the next day, 90s fired on enemy bunkers, observation posts, and trenches in the vicinity of Siberia and Stromboli. During 11 August, two gun tanks blasted at targets immediately beyond Siberia and others to the west of that outpost.

Towards the end of the 11th, the critical part of Bunker battle began for the tankers also. Those elements of Lieutenant Colonel

[38] Many of these targets were CCF choke points, dumps, and weapons emplacements. Targets were identified and confirmed by a highly developed system that employed air spotting, aerial photographic interpretation, artillery evaluation, and POW interrogation.

John I. Williamson's battalion supporting the diversion and the subsequent main attack pulled into positions south of Hill 122 on the MLR and to the right in the Company F sector. It was not until the next day that the tanks operating with the 1st Marines reached a peak in gun support for the Bunker fight. Beginning with the defense of Hill 122 from 1600 that day, and for the next 26 hours, the tankers placed 817 shells on targets effecting the Chinese capability of capturing Bunker and Stromboli. In addition to the heavy ammunition, the Company C tanks, augmented by the 1st Marines antitank platoon and five tanks from the division tank reserve, fired 32,000 rounds of .30 caliber machine gun ammunition.

Except on the 11th, most of the tank firing in the fight for Bunker Hill through 14 August was accomplished during the hours of darkness. On the latter date, the cannons and machine guns of the mediums blasted directly at Chinese outposts opposite Colonel Layer's regiment. The number of rounds that day fell off considerably from the high on the 13th; on the 15th the tanks in the 1st Marines area did not fire at all. Heavy rain that had accompanied the late afternoon thundershower that day made movement forward to firing positions impractical. By the next day, however, the ground was solid enough to permit some maneuvering by the tracked vehicles. They fired 52 rounds of 90mm shells and 14,750 machine gun rounds at automatic weapons positions and bunkers on the western slope of Hill 122. This marked the final tank mission in support of the 1st Marines in the battle for Bunker Hill.

During the early part of the August fighting, tanks of the division were able to get the first real test of a technique of night support,[39] and at the same time experiment with a towing device to permit retrieval of disabled vehicles under fire without getting outside the tank. The use of the lights to support both the diversionary force and the defense of Hill 122 showed the value of these instruments. Lieutenant Colonel Williamson recommended that tanks be employed in pairs, one to spot and adjust fire and the other to fire. With respect to the towing device, he considered the new piece of equipment an improvement over the manual hook-up method, but noted that the device limited tank maneuverability and had a tend-

[39] The use of fighting lights to illuminate targets for tank gunners had been undertaken in July, but the results were inconclusive, owing to failure of one of the bulbs of the two lights tested. 1st TkBn ComdD, Aug 52, App. VI, Encl. 2. Declared the G-3, 1stMarDiv: "The diversion on Siberia was 100 percent effective, due largely to the new tank battle lights which we were using for the first time." *Honsowetz ltr II.*

ency when bouncing up and down over rough terrain to dig into the ground, impeding the forward progress of the vehicle.

In Retrospect[40]

Whether the sacrifice of Siberia in favor of the seizure of Bunker justified the outcome can be determined, in part, by looking back to the division commander's reasons for this decision. He had cited three advantages in seizing and occupying Hill 122 instead of 58A. One, tactical surprise achieved by an attack on the former, was an unqualified success. That Bunker Hill would provide more defensible terrain and at the same time add strength to the main line were two sound judgments that the test of time would bear out. The third point, that observation into the enemy's outpost line would be increased from the higher hill, also proved to be correct.

Only the inability to neutralize Hill 58A effectively from Bunker cast any doubt on the considerations. At night the enemy could occupy Siberia both for firing positions and flank security to attack friendly forces moving down the corridor east of Hill 122. Action to counter these two enemy actions came mainly from MLR forces.

One measure of the results of the Bunker Hill fighting is seen in the price paid. Chinese losses were estimated by the 1st Marine Division at approximately 3,200, including more than 400 known dead. Marine casualties in the action were 48 killed and 313 seriously wounded. Several hundred additional wounded were treated at 1st Marines medical facilities and returned to duty shortly thereafter.

To replace combat losses in the infantry regiment, General Selden on 12 August directed that rear area service and support units fill the vacancies. Two hundred Marines, nearly all of them volunteers, were provided to Colonel Layer by the 14th. To offset other losses within the division, its commander similarly had requested on 12 August that the Commandant, General Lemuel C. Shepherd, Jr., authorize an air-lifting of 500 enlisted Marine infantrymen to the 1st Marine Division as soon as possible. Pointing out that mounting battle casualties had reduced the effective strength of the

---

[40] Unless otherwise noted, the material in this section is derived from: FMFPac, *1st-MarDiv Sum, Jul–Oct 52; PacFlt EvalRpt* No. 5, Chaps. 8, 9; 1stMarDiv, 1stMar ComdDs, Aug 52.

division, General Selden also urged that each of the next two monthly replacement drafts scheduled for the division be increased by 500 more enlisted men. After some debate at the next senior administrative headquarters,[41] the request was granted by General Shepherd, and the emergency replacements were made available from the 3d Marine Division at Camp Pendleton, California. The initial replacement of 500 Marines arrived on 21 August.

More men to replace divisional combat losses might have been required had not the medical support been such an efficient operation. After the battle, the regimental surgeon, Lieutenant Robert E. Murto, called for a review of the medical facilities in effect during the Bunker, Siberia, and Stromboli fighting. In attendance were the battalion doctors and the division surgeon, Captain Lawrence E. Bach. Participants discussed both the major difficulties and routine procedures involved in medical care of the wounded. Problem areas were the high incidence of heat exhaustion, ground transportation of the wounded, enemy artillery fire that interfered with helicopter evacuations, and the need for increased medical support under battle conditions.

Regarding the last category, the surgeons noted that medical supplies during the heavy fighting of 9–16 August were never at a dangerously low level. The only shortage that had developed was in stretchers, due to the normal delay in transfer of stretchers from medical stations along the evacuation route to the company forward medical facilities. To help combat the Chinese artillery problem, medical officers had placed aid stations on the reverse slopes of hills. There was no available or known solution to hastening and easing the movement of battlefield casualties over the ground. The armored personnel carrier offered some protection from ground fire and a ride less painful than one in a truck, but the wheeled vehicles remained the most widely used.

There was little that could be done about the number of heat exhaustion cases. High temperature and humidity, vigorous activity,

---

[41] CG, FMFPac, Lieutenant General Hart, requested the Commandant to delay decision until FMFPac could survey the combat replacement situation and aircraft availability. After a quick evaluation of both these factors, General Hart on the 14th recommended approval. FMFPac ComdD, Aug 52, App. I, Encl. (6). The air lift of 500 replacements to Korea was an "all out effort for Marine Aviation Transport based on the West Coast. This general support of Korean based forces demonstrated the total capability of Marine Aviation in support of ground forces." MajGen Samuel S. Jack to Hd, HistBr, G-3 Div, HQMC, dtd 27 Jun 67, hereafter *Jack ltr.*

and the wearing of the armored vest (and to some degree, the steel helmet), combined to produce the casualties. All the surgeons agreed that regardless of the number of heat casualties, the wearing of these two items must continue. Regimental doctors credited the armored vest with saving the lives of 17 Marines. Several other Marines, they noted, had received only slight head wounds from bullets that had spent most of their velocity penetrating the steel helmet.

Helicopter evacuation saved the lives of other Marines. The doctors credited the flying skills and bravery of the evacuation pilots for these rescues. Immediate response to day and night calls was instrumental in the recovery of numerous Marines. Rear Admiral Lamont Pugh, Surgeon General of the U.S. Navy, commented upon the value of the helicopter and on other reasons for success of medical support. After a Far East inspection trip, which included a visit to the 1st Marine Division during the battle of Bunker Hill, Admiral Pugh expressed the following opinion:

> . . . [I] attributed the new low record "2% mortality" of those men wounded in action to the bullet resistant vest, to skillful frontline surgery with availability of whole blood, the utilization of helicopters for casualty evacuation direct to hospital ships and rear area hospitals, and the efficient manner in which the Hospital Corpsmen of the Navy fulfilled their mission with the Marines.[42]

In another logistical area, the performance was not quite as satisfactory, for the level of supply of one important item—illuminating shells—fell dangerously low during the Bunker fighting. On 16 August, 3/1 reported early in the morning that "artillery illumination was exhausted and 81mm mortar illumination was fast diminishing."[43] To replace the shell-produced light, the regiment used a flare plane.[44]

Ammunition supply appeared to be no problem to the Chinese. The rate and frequency of mortar and artillery fire proved that the enemy had a vast store of these shells. During the heavy fighting, the division observed that the enemy expended approximately 17,000 mortar and artillery rounds in the 11–16 August period of

---

[42] *PacFlt EvalRpt* No. 5, Chap. 12, p. 12–8.

[43] 3/1 ComdD, Aug 52, p. 4.

[44] Earlier, on 13 August, a flare drop requested by the 1st Marines went awry when the aircraft got off course and dropped the flares forward of the 5th Marines main line. 1stMarDiv G–3 Jnl, 13 Aug 52.

the battle. It was noted for the first time that the Chinese used mortars primarily in support of limited attacks.

About the enemy's reliance on mortars and the technique of their employment, the 1st Marine Division reported:

> This was particularly true of his 60 and 82mm mortars, which are easily displaced forward and shifted to alternate positions. These light mortars were difficult to locate by our observers mainly because of the small size and limited development of their positions, and the fact that they are moved frequently. A large number of enemy mortars were fired from bunkers deep in the ground with only a narrow aperture at the top through which to fire. There were some instances, during the Battle of Bunker Hill, when the enemy brought his 60mm mortars out from cover on the forward slope and set them up in the open near the crest of the ridge. After delivering several rounds, the mortars would then displace quickly back to a covered position. During August, mortar fire averaged between 50 and 60 percent of the total incoming received by the 1st Marine Division.[45]

Further information about the Chinese was also derived at this time, although not always directly associated with the battle. Deserters picked up in the left sectors of the 1st and 5th Marines on 12 and 13 August and papers taken from enemy dead on the 13th confirmed earlier-reported dispositions of Chinese units. One prisoner, from the artillery regiment of the 118th Division, the unit facing the major part of the 1st Marines line, indicated that another artillery regiment had been assigned to support his division. If true, this extra unit would account for both the increased Chinese fires in the Bunker area and the additional artillery emplacements that photo planes had spotted in the 118th Division sector. Infantry units of this division, the Marines observed, introduced no new techniques or equipment during the battle. Prior intelligence had provided the 1st Marines with typical enemy ground attack tactics. Neither the Chinese envelopment of Siberia, Stromboli, and Bunker nor the diversion against Hill 48A before the main attack on Hill 122 represented a departure from normal CCF practice.

Nor was the earlier Marine diversion new, but unlike the Chinese attempt, the 1st Marines tactic was successful. Just before the maneuver, the division pulled off another strategem, described by General Selden in a letter to General Shepherd:

> I worked a ruse that morning which proved to be very profitable. Throughout the Eighth Army front, it had been routine to put on a strike,

[45] 1stMarDiv ComdD, Aug 52, p. 2.

> this to be followed by smoke, then a good artillery barrage, with troops following for the assault. This was done with the exception that there were no troops. The enemy, thinking that there were troops, opened up with everything. The only damage inflicted was on their own forces . . . While they were firing on their own troops, we again opened fire with our artillery, just to help the situation along.[46]

One technique the Marines employed in the Bunker Hill battle was defense of the reverse (protected) side of the hill. Although counter to the usual American military practice, the reverse slope defense was required by the intense artillery and mortar fire massed upon the front slope defenders. As the 3/1 battalion commander later commented:

> It's true, we suffered from the heavy incoming—but had we had to work replacements, casualties, and supplies all the way up to the (forward) military crest of Bunker—the losses would have been prohibitive. With the weight of the incoming and our inability to get greater infantry mass onto the battlefield at one time, a conventional defense would have been far more costly . . . [after] the damage done to Baker Company in the [12 August] afternoon attack . . . had we not gone into a reverse slope defense, we could not [have held] with the strength at hand."[47]

On the other hand, a tactical weakness of the reverse slope defense, that "plagued us until the end of the battle,"[48] was the fact that the 1st Marines initial gain was not more fully exploited. As the battalion commander explained:

> To be successful, in a reverse slope defense, the defender must immediately counterattack, retake and reoccupy the *forward* slope of the position as soon as enemy pressure diminishes. Because of the incoming and primarily because of our overextension in regiment, we . . . [employed] piecemeal commitment . . . and fed units into the battle by company, where we should have employed our entire battalion in counterattacks to punish the withdrawing force and restore the forward slope. To the very end, lack of decisive strength prevented this. We stayed on the reverse slope all the way, except for brief forays to the forward slope.[49]

Some officers felt, in retrospect, that a more feasible solution during the August battle might have been to move all three battalions on line—3/1, 1/1, and 2/1, with the reserve battalion

---

[46] MajGen John T. Selden ltr to Gen Lemuel C. Shepherd, Jr., dtd 14 Aug 52.
[47] *Armitage ltr* and draft MS comments, p. 7. For further details of the Bunker Hill action, see *Armitage ltr* in v. V, Korean comment file.
[48] *Ibid.*, p. 8.
[49] *Ibid.*

(1/1) deployed on a narrow front. This would have provided decisive strength on Bunker and the MLR behind it to give greater depth counterattack capability, and better control at the point where needed.[50] Departure from standard doctrine by employment of the reverse slope defense furthered the existing controversy as to the best method of ground organization in the division sector. But it was to be some months before a change would be effected.[51]

Tank, artillery, air, and ground Marines participating in the battle of Bunker Hill gave up one outpost but took another, one that added strength not only to the outpost defense but also to the main line. A well thought-out plan and its skillful execution permitted Marines to take the critical terrain quickly without crippling casualties. Defense of the position on Hill 122 was complicated not so much by the Chinese infantry action but by the intensive mortar and artillery shelling. The Marines' capability to defend was enhanced by close coordination among artillery, air, and tank units. Chinese casualties, by estimate, were 500 percent more than the losses actually suffered by the Marines. The battle of Bunker Hill resulted in the first major Marine action and victory in West Korea. It ushered in two straight months of hard fighting, the most difficult ones yet for Marines on the western front.

---

[50] *Ibid.*, p. 9.

[51] As the military situation changed in Korea to become increasingly one of a battle of position and attrition, the Marine Corps Basic School, Quantico, Va. curriculum was revised to give greater emphasis to tactics of positional warfare. Close attention was paid to terrain evaluation, employment of infantry units, offensive and defensive use of automatic and supporting weapons, night counterattacks, field problems of reverse slope defense, and even tasks of "research into WW I—and the American Civil and Revolutionary Wars for the tactic of Reverse Slope defense." *Armitage ltr.*

CHAPTER IV

Outpost Fighting Expanded

From the Center Sector to the Right—Early September Outpost Clashes—Korean COPs Hit Again—More Enemy Assaults in Late September—Chinese Intensify Their Outpost Attacks—More PRESSURE, More CAS, More Accomplishments—Rockets, Resupply, and Radios

From the Center Sector to the Right[1]

FOLLOWING THE progressively faltering Chinese attacks against Bunker Hill in mid-August, the 1st Marines in the center MLR sector witnessed a period of decreased enemy activity. By sun-up on the 17th, Captain Ksycewski's Company C, from Lieutenant Colonel King's 1st Battalion had relieved B/1/1, marking the second complete tour of duty at Hill 122 for Company B that month. In two days on the shell-torn crest, Company C received only a single enemy probe and only a few rounds of artillery and mortar fire. In the early morning hours of the 19th, D/2/1 assumed responsibility for Bunker and Hill 124. These new occupants of the disputed property almost immediately were subjected to larger and more frequent Chinese probes as well as increased fire from CCF supporting weapons.

Enemy ground action was directed against the Marine flank, especially the right. Four Chinese infantrymen attempted to infiltrate this corner of the Bunker Hill defenses just before sunrise on 23 August. One even made his way to the top of Hill 122 where he fired downhill at several Marine defenders, wounding one. A moment later this lone Chinese's reconnaissance efforts was rewarded by a fatal hit from a Marine sniper's rifle.

---

[1] Unless otherwise noted, the material in this section is derived from: 1stMarDiv ComdD, Aug 52; 1stMarDiv PIRs 661–675, dtd 18–31 Aug 52; IstMar, 5thMar, 2/1, 3/1 ComdDs, Aug 52.

Captain Moody's Company F next took over the two-hill complex. That night, the 24th, the Chinese shelled the two hills and probed their defenses but again showed no inclination to press an attack. On the following night, however, the Chinese became more aggressive. At dusk, two squads charged the right flank of Bunker Hill, threw hand grenades, and fired their submachine guns briefly at the Marines. The enemy then retired, but about an hour afterwards, a force estimated at two-company strength assaulted the outpost defenses from the center to the right. At the same time, enemy shells began exploding around these Marine positions. Captain Moody called for artillery and tank fire on the attackers. Pushing forward, the Communist infantrymen forced a small opening in the defense perimeter; by this time, a standby platoon on the MLR was moving forward to strengthen the Bunker garrison. Upon arrival of the Marine reinforcements, at midnight, the Chinese soldiers withdrew. Simultaneously, the incoming artillery and mortar fire diminished, and in less than a half hour all firing had ceased.

After the enemy had pulled back, Company F sent its platoon out to reoccupy a forward listening position temporarily abandoned during the second attack. Chinese soldiers immediately contested this advance and, after a local fire fight, caused the Marines to retire once more. That action ended the significant Bunker Hill action in August. In the spirited infantry fighting and artillery dueling during the night of 25–26 August, Marines suffered 65 casualties, including 8 killed. The Chinese losses were estimated at 100 killed and 170 wounded. Supporting arms fire had contributed largely to the high casualty figures on both sides.

During August, whenever a lull had occurred in Colonel Layer's 1st Marines embattled sector, it almost invariably signaled a step-up of Chinese action elsewhere along the 1st Marine Division MLR. When frustrated in their attacks against the positions held by the 1st Marines, the enemy invariably turned his attention to the right of the line, manned since June by the 5th Marines. During August the Chinese seized three outposts forward of the 2/5[2] right battalion line, which it had been the Marine practice to man during daylight hours only. The trio, forming a diagonal line southwest to northeast, in front of the battalion sector were Elmer, Hilda, and Irene.

---

[2] Command responsibility for this sector changed on 20 August, when Lieutenant Colonel William S. McLaughlin took over the battalion from Lieutenant Colonel Cross.

After dusk on 6 August the enemy had advanced to COP Elmer, on the far southwest end, and by skillful coordination of their infantry and supporting fires denied the position to the Marines approaching to reoccupy the outpost early the next morning. An hour before midnight on 11 August, another 2/5 patrol had attempted to temporarily occupy Hilda, in the center, during the diversionary fires supporting the Bunker Hill attack. As the Marines neared the outpost, however, they discovered the Chinese had already occupied it. Enemy mortar and artillery fire drove the patrol back to its own lines.

A similar situation occurred at dawn on 17 August, when the Marine outpost detail moved forward to occupy Irene during daylight hours and found the Chinese already on the position. Enemy troops fired at the Marines, pinning them down.[3] Although two rescue units were dispatched to support the Marines, CCF fire interdicted their route of approach. When it became evident the second reinforcement party could not reach its objective, the outpost detail was ordered to pull back to the MLR. The Chinese continued to occupy Irene, the last outpost lost in August, for the remainder of the 2/5 tour on line.

For the remainder of August the Chinese were apparently content to hold what they had gained without immediately seeking additional positions. As a result, operations along the front were mostly limited to patrol action. Chinese infantry units, usually no larger than a squad, regularly fired on Marine patrols, engaging them for a short period from afar, and then quickly breaking off the contact. Seldom was this small unit action supported by artillery or mortars.

On two occasions late in the month, however, the Chinese showed more spirit. Both encounters took place during the early evening hours of 22 August when Chinese patrols came upon two different Company F ambushes operating forward of the 2/5 sector. Heavy casualties were suffered by both sides.

The next day a brief but heavy period of rainfall began with nine inches recorded between 23–25 August. Although the flooding conditions in the division sector were not so extensive as the July rains, they curtailed ground activity considerably and air action to a lesser

---

[3] To escape the murderous hostile fire, the Marines sought shelter in a trench nearby. During the ensuing clash, a Chinese grenade landed in the midst of the Marines. Private First Class Robert E. Simanek, E/2/5, unhesitatingly threw himself upon the deadly missile an instant before it exploded. Although gravely wounded, his courageous action prevented injury or death to fellow patrol members. The following year, President Dwight D. Eisenhower presented the Medal of Honor to the Detroit, Michigan Marine for his "daring initiative and great personal valor."

degree. Division roads were badly damaged but not trenches and bunkers, strengthened as a result of the experience with the July floods. High waters made the ferry inoperable at the Honker Bridge site and also washed out Widgeon Bridge, where the Imjin crested to 42.5 feet. If the sudden flash floods wreaked havoc with some of the Marine division installations, the Chinese were the recipients of similar disfavors; intelligence indicated that damage to the CCF frontline positions was even more severe than to the JAMESTOWN defenses.[4]

The end of August saw the relief of General Selden as Commanding General, 1st Marine Division. He was succeeded on the 29th by Major General Edwin A. Pollock. A brief ceremony at division headquarters, attended by senior officers of EUSAK and KMC, marked the event. Earlier that month, in recognition of his services to the Korean defense, President of the Republic of Korea, Syngman Rhee, had awarded General Selden the Order of Military Merit, Taiguk, the highest Korean award.

The new division commander, General Pollock[5] had commanded the 2d Marine Division at Camp Lejeune, North Carolina just prior to his Korean tour. He had more than 30 years of military experience. During World War II, he had participated in no fewer than five major campaigns in the Pacific, including the first at Guadalcanal, where he earned a Navy Cross, and one of the war's most costly battles, Iwo Jima. Following the war, he had served at Marine Corps Schools, Quantico, in command and staff assignments, and later at Headquarters Marine Corps where in July 1949, he had received his first star.

Early September Outpost Clashes[6]

The new division commander shortly received a first-hand demonstration of the ferocity and persistence of the Chinese Communists opposite his division. On 4 September, the enemy suddenly stepped up his activities which had recently been limited to sporadic probes

[4] 1stMarDiv PIR 669, dtd 25 Aug 52.

[5] DivInfo, HQMC, Biography of MajGen Edwin A. Pollock, Jan 56, rev.

[6] Unless otherwise noted, the material in this section is derived from: 1stMarDiv, 1stMar, 5thMar, 2/1, 3/1, 2/5, 3/5 ComdDs, Sep 52; KMC Regt UnitRpts 188–189, dtd 6–7 Sep 52.

and occasional artillery fire against Bunker Hill. At 2030 that date Captain Moak, E/2/1, commanding officer at the Bunker outpost, reported that an artillery preparation was falling on his positions. Ten minutes later he radioed 3/1[7] that an enemy platoon was vigorously probing his right flank. When Company E Marines returned a heavy volume of small arms fire, the enemy retired.

This Chinese withdrawal was only temporary, for the initial probe proved the forerunner of more serious activity. Again at 0100 on 5 September a heavy deluge of Chinese mortar and artillery began raining on Hill 122. The intense preparation had apparently convinced the Chinese attacking force that they had eliminated resistance at the Marine outpost, for their soldiers walked upright toward Marine positions, without bothering to make any attempts at concealment. After discovering that a stout defense was still being maintained at Bunker, the Chinese again withdrew and reorganized.

When they resumed the attack, the Chinese used considerably greater caution. This time, in addition to small arms, automatic weapons fire, and a hail of grenades, their assault was supported by artillery and mortars. The results of this concerted effort were not too rewarding, however. Assaults on the center of Hill 122 were repulsed and attempts to crack the left perimeter of Company E's defenses were even more speedily beaten back. A number of Chinese attempting to outflank the E/2/1 defenders inadvertently strayed too far to the right of the outpost and found themselves advancing against the MLR south of Hill 122.

When JAMESTOWN forces engaged these wanderers by fire, the latter quickly realized their mistake and wheeled left for a hasty retreat. They immediately came under fire of their own troops, some of whom had meanwhile penetrated 60 yards into the extreme right of the Bunker positions. At this point, Captain Moak's Company E launched a counterattack and restored its positions on the right. This action forced a general withdrawal of the Chinese force, which the Marines estimated at battalion strength. Lieutenant Colonel Sidney J. Altman[8] subsequently advised division headquarters that his men had killed 30 enemy soldiers and estimated that as

---

[7] Normally a component of the 2d Battalion, Company E had been attached to the 3d Battalion on 1 September when the company took over the Bunker Hill outpost. The relieved Company H was then attached to 2/1, the reserve battalion, from 1–3 September.

[8] On 20 August Lieutenant Colonel Altman became the commander of 3/1 in relief of Lieutenant Colonel Armitage.

many as 305 were probably wounded. This high rate of casualties was attributed, in part, to the enemy's mistaken sense of direction, their direct walking approach which had made them easy standing targets, and to the box-me-in artillery fires supporting the defenders. Marine losses were 12 killed and 40 wounded, caused mostly by Chinese mortars and artillery.

Although the left battalion area was the center of attention in the 1st Marines line early on 5 September, the far right sector was not entirely neglected either. Five minutes after their initial attack on Bunker, other Chinese units also lunged at the Hill 48A outpost, Stromboli. An estimated reinforced platoon, supported by three active machine guns on Hill 104, 850 yards to the north, employed submachine guns, rifles, and grenades in their attack. This battle lasted for nearly two hours, until the Chinese soldiers withdrew at 0240. There were no Marine losses. No tally or estimate was made on the number of enemy KIA or WIA. It was presumed that some of the Communists did become casualties since the three machine guns that had been chattering away to support the attacker's ground action suddenly went silent after Marines called down mortar and artillery fire on the Hill 104 positions.

The probes of 1st Marines positions at Bunker Hill and, to a lesser degree, at Stromboli were repeated in the 5th Marines right regimental sector. At almost exactly the same time Colonel Eustace R. Smoak's regiment[9] was struck at five of its forward outposts. In the case of OP Gary, on the right, the enemy merely shelled the position for 40 minutes. Against the four other outposts, known as Allen, Bruce, Clarence, and Felix, the Chinese employed both fire and asault troops. (See Map 12.) At Felix the action had begun at 0130, a half hour later than at the adjacent outposts. The difference was probably due to a C/1/5 ambush[10] which had engaged an enemy force operating between Donald and Felix. After a brief five minute fire fight the Marines broke off the action, pulling back to Felix. The other three outposts, clustered to the left of the 3/5 sector, received the brunt of the enemy thrust which lasted for an hour and 20 minutes before the Communists withdrew.

---

[9] Colonel Smoak had relieved Colonel Culhane on 15 August.

[10] Although 1/5 (Lieutenant Colonel Alexander W. Gentleman) was the regimental reserve at this time, the regiment had assigned one company to 2/5, manning the right sector.

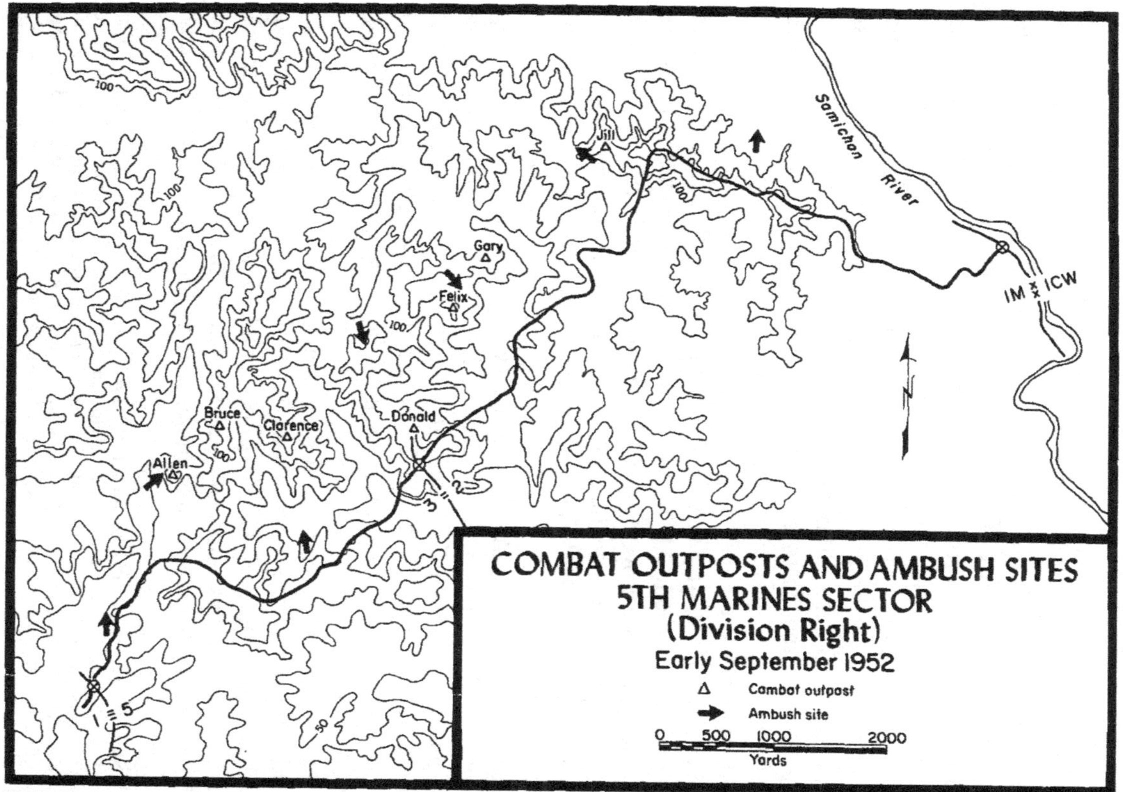

MAP 12

K WHITE

Employing a squad against both Allen and Clarence, and sending a reinforced company against Bruce, the enemy alternately assaulted and shelled the positions until 0420, after which the Communist units policed the battlefield for casualties and withdrew to the north.

Although there was no official estimate of enemy losses, one Marine at outpost Bruce was credited with inflicting approximately 200 casualties by fire from two machine guns, a carbine, and grenades. He was Private First Class Alford L. McLaughlin, of I/3/5, who was later to receive the Medal of Honor for "conspicuous gallantry and intrepidity." Another Marine from the same company was posthumously awarded the medal. Private First Class Fernando L. Garcia, although gravely wounded, had thrown himself on a hostile grenade to save the life of his platoon sergeant during the Chinese rush to take OP Bruce.

At daybreak the I/3/5 defenders at Bruce, commanded by Captain Edward Y. Holt, Jr., were confronted by an almost unbelievable scene of destruction. All of the bunkers on the forward side of the hill had been destroyed by Chinese mortar and artillery; on the reverse slope, only two had escaped ruination. Marine losses were 32 dead and wounded.[11] To restore the position the 3/5 commander, Lieutenant Colonel Oscar T. Jensen, Jr., directed replacements forward immediately. Carrying emergency supplies, including building materials, the relief element reached Bruce about 1000. Evacuation of casualties was the first task and at 1045 the relieved detail was on its way back to the MLR. Later that day a supply party reached the outpost, having been temporarily delayed by Chinese interdicting fire.

Reinforcement of Bruce and the repair of its defenses were considerably slowed by the continuous rain of enemy projectiles during daylight. Marine and USAF pilots bombed and napalmed enemy bunkers and troops north of JAMESTOWN in the 5th Marines sector. Ten air strikes were executed in support of the 5th Marines that day.

---

[11] Still another award of the Medal of Honor was to come out of the action that ended on 5 September. Hospitalman Third Class Edward C. Benfold had ministered aid to several wounded Marines and was searching for others who needed medical attention when he saw two wounded Marines in a shell crater. Just as he neared its edge two grenades fell into it and two Chinese prepared to assault the Marines. "Picking up a grenade in each hand, Benfold leaped out of the crater and hurled himself against the onrushing hostile soldiers, pushing the grenades against their chests and killing both. . . . He gallantly gave his life for his country." Medal of Honor citation, case of Hospital Corpsman Third Class Edward C. Benfold, USN, 4168234.

Early on 6 September, 10 minutes after midnight, long-range machine gun fire, buttressed by artillery and mortars, hit outpost Bruce. After 35 minutes the firing subsided, but again at 0305 the outpost experienced a heavy rate of incoming. At about this time, the Communist soldiers massed for an assault on the battered position. Marine defenders called down the artillery box, and the Chinese dispersed.

That evening, at 1915, the outpost commander reported that the Chinese had again resumed a steady shelling of the position. The bombardment continued for an hour. After these heavy preparatory fires, a wave of enemy infantry began scrambling up the sides of Bruce. At the same time, outpost Allen to the left came under long-range fire from enemy strongholds to the west and north. After the Chinese made their initial rush against Bruce, a second and third attack fared no better. Each was met and repulsed by the 5th Marines.

After the third abortive attack, a period of deathly stillness descended upon the contested hill. Occasionally, an enemy mortar round found its mark among the scattered, splintered bunker timbers and the caved-in trenches, which connected the sandbag and lumber positions. At 0145 on the 7th, the Chinese interrupted the uneasy peace that had settled upon Bruce with a brief, heavy preparatory fire.

Exactly an hour later, an estimated two Chinese companies advanced up the forward slopes, using demolitions to destroy any friendly bunkers their artillery and mortar had not earlier completely wrecked. By the time this newest assault had raged for 30 minutes, nearly every 3/5 defender had become a casualty. Still the Marines refused to give ground, dealing first with the forward slope assault by the Chinese and later with those who attempted to envelop the Marines on the reverse side. On the MLR Marines first observed enemy flares falling between outpost Bruce and Line JAMESTOWN. Soon thereafter the Chinese policed the battlefield. By 0400 the Communists retired, and the fight for this key outpost had ended in failure.

During the 51-hour siege of Outpost Bruce, 19 Marines had been killed and 38 wounded. At the adjacent 5th Marines outposts, additional losses were 5 killed and 32 wounded. More than 200 enemy dead were counted. During the last eight hours of the vicious, close-in fighting at Bruce, it was estimated that another 200 Chinese had been wounded.

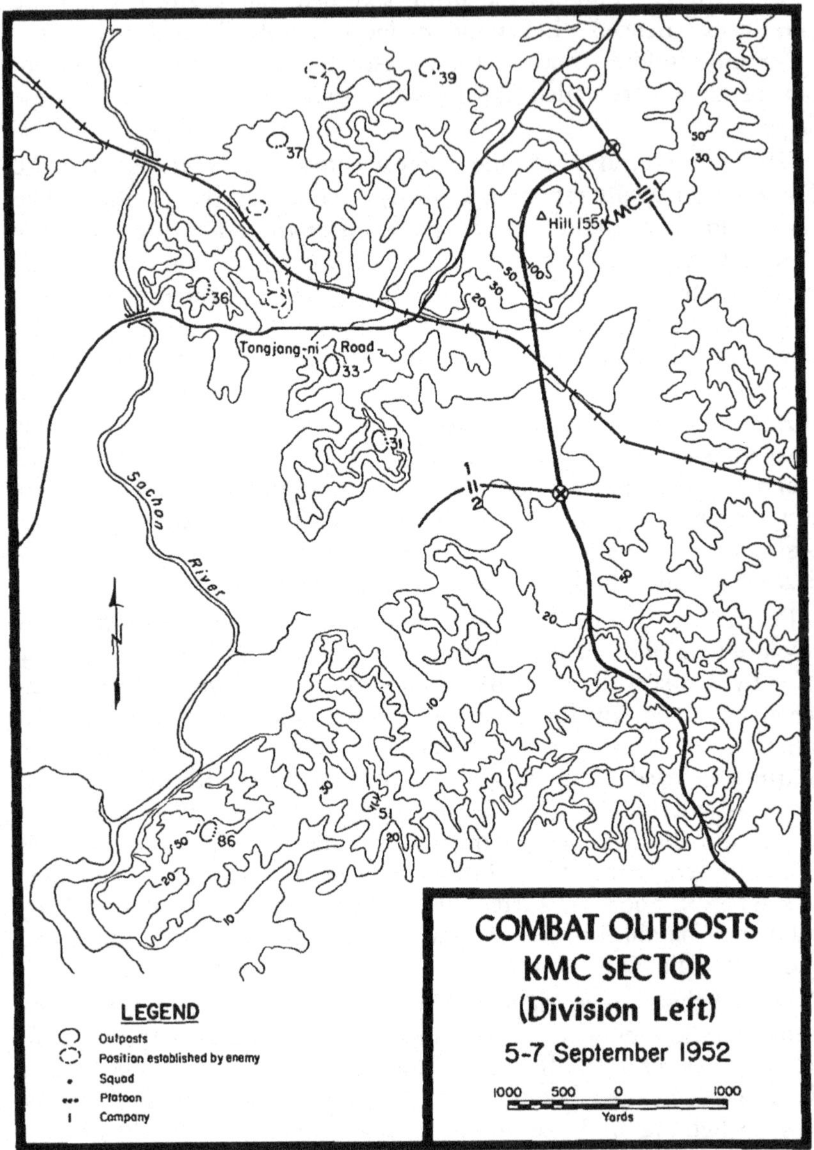

MAP 13

K.White

The Korean Marines, holding down the western flank of the three mainland regimental sectors in the 1st Marine Division line, also received a share of the enemy's attention. At dusk on 5 September, Chinese barrages began to smash Outpost 37,[12] the first of a trio of positions that would merit hostile attention for the next 22 hours. Throughout the following day the Chinese continued their mortar and artillery fire against Outposts 37 and 36, and the regimental observation post located on Hill 155 (also called Hill 167) to the rear of the MLR. (See Map 13.) The heaviest enemy fire was directed against OP 36, a small rise in the low land terrain midway between the Sachon River, on the west and the Munsan-ni-Kaesong rail line, 600 yards to the east.

At 1605 a 50-round barrage struck OP 36. After this harassing fire there was a lull until 1810 when Chinese artillery and mortars again resumed a steady pounding of the three positions. One hour later enemy soldiers hit both outposts. Twice the attacking company assaulted OP 37 but neither effort represented, in the view of the defenders, a serious attempt at capture. Less than a mile south at OP 36, however, the enemy motive appeared to be quite different.

Crossing the Sachon just north of the Freedom Gate Bridge (also known as the highway bridge), the Communist infantry moved to assault positions on the west, north, and northeast sides of the outpost. At 1910, the Chinese began their first rush. It was repulsed, as was a second one. Another artillery barrage, joined this time by tank fire, preceded the third attempt. At this point communications went out at the besieged outpost. At 2150, a squad leader from OP 36 reached the 10th Company CP to report that his position had fallen. In 30 minutes a communications link was re-established with the outpost. The defending Koreans reported that although enemy troops had overrun much of the hill, they had subsequently withdrawn, apparently because their losses had been so heavy.

Casualties and damage were severe. The Korean regiment estimated that 110 enemy had been killed or wounded. An early morning KMC reconnaissance patrol counted 33 dead Chinese in the vicinity of OP 36. The attacking force had also left behind much equipment, including more than 100 grenades and several automatic

---

[12] Contemporary records of the 1st KMC Regiment for 1952–1953 identify this as Outpost 37. Current reviewer comments refer to this hill as OP 67. LtCol Kim Yong Kyu, ROKMC, ltr to CMC, HQMC, dtd 5 Jul 67.

weapons. No papers were found on the dead Communist soldiers, but many propaganda leaflets had been dropped around the outpost. Korean Marine losses at OP 36 were nine killed and seven wounded. At OP 37 there were four casualties; at the regimental CP, one Korean and two U.S. Marines had been killed by enemy artillery. Chinese incoming, estimated at 2,500 rounds during the two actions, had also caused major damage to part of the OP 36 defenses, but inflicted less harm to the other two positions. Repairs were begun before daylight.

Korean COPS Hit Again[13]

After the stepped-up enemy ground activity in early September, both Chinese and Marine frontline units resumed their earlier pattern of combat patrols, probes, and ambushes. Possession of Bunker Hill remained the immediate objective of the enemy and his activities in the middle of the Marine line were directed to this goal. Once again on 9 September a marauding Chinese platoon, employing grenades and submachine guns, sounded out the Bunker defenses, now manned by G/3/1 (Captain William F. Whitbeck, Jr.). After a tentative investigation, the enemy withdrew. That same day, expanded patrol and raiding activities were undertaken by Marine line battalions.

These sharply increased offensive measures resulted, in part, from the Communist interest, as evinced during the summer truce negotiations, in certain forward positions held by UNC units. On 7 September, the CG, I Corps had alerted his division commanders to the fact that the enemy "may attempt to seize and hold certain key terrain features . . . over which there was extensive disagreement during [the 1952 summer truce] negotiations for the present line of demarcation."[14] Since much of the critical land was in his sector, Major General Kendall further warned his division commanders "to take the necessary action within your means to hold all terrain now occupied by your divisions."[15] Critical terrain features in the 1st Marine Division area of responsibility were Bunker Hill and the

---

[13] Unless otherwise noted, the material in this section is derived from: 1stMarDiv ComdD, Sep 52; KMC Regt UnitRpts 195–202, dtd 13–20 Sep 52.
[14] 1st MarDiv ComdD, Sep 52, App. I, # 8.
[15] *Ibid.*

height on which COP Bruce had been established (Hill 148), in the center and right regimental sectors respectively.

Two days later, General Pollock amplified this directive by underscoring the necessity for holding these two positions, plus eight more he considered vital for sound tactical defense. These additional positions, from west to east, were Hills 86 and 37 in the KMC sector; Hills 56 and 48A in the center sector; and the outposts then known as Allen, Clarence, Felix, and Jill, all the responsibility of the right regiment.[16]

Although the eastern part of the division main line thus contained at this time more key hills than any other Marine sector, much of the increase in Marine patrol and ambush activity took place in No-Man's-Land forward of the middle frontline regiment. Of the two JAMESTOWN sectors manned by U.S. Marines, the one in the center of the division area offered better ground for infantry operations.

On the divisional western flank, the Korean Marines conducted frequent infantry-tank patrols during the second and third weeks of September, but the enemy opposite the KMCs initiated little ground activity. Instead, the Chinese relied upon their supporting weapons to provide the contact. For a seven-day period ending 19 September, a total of 2,375 enemy rounds had fallen in that regimental sector, an average of 339 per day. Nearly a third had been in the vicinity of Hill 36.

Before sunrise on the 19th, a Chinese infantry company had crossed the Sachon in the vicinity of the railroad bridge. Once on the east side, the enemy soldiers concealed themselves in caves and holes, remaining there until dusk. Then, when they came out of hiding, the Communists held a briefing and organized themselves into three attack groups. As these advance infantry elements approached their objective, OP 36, other reinforcing units were prepared to seize OP 37, to the east, and OPs 33 and 31, to the south. Artillery and mortar preparation supported these diversionary attacks.

The main assault was accompanied by even heavier shelling. As the three assault units reached the bottom of the hill at OP 36,

---

[16] When the 7th Marines took over this sector from the 5th in early September, the names changed to Carson, Vegas, Detroit, and Seattle respectively. COP Bruce was also redesignated as Reno. Since the old names of the outposts were well known to the enemy, for purposes of security it was decided to identify them differently. U.S. cities were selected.

artillery, mortars, and tanks had fired more than 400 rounds. Approaching from the north, east, and west, the Chinese scrambled up the hill, gaining control of the wrecked defenses by 2000. Sporadic exchanges of fire lasted until nearly midnight. At 0115 the Korean Marines attempted to retake the hill. The counterattack was cut short, however, upon discovery of another enemy unit moving towards the outpost and then only one-half mile away. Three hours later the enemy came back in strength when a CCF platoon successfully overthrew the outpost at 0520. This new assault occurred without any warning and was so swiftly executed that a number of the KMC defenders found themselves encircled and trapped at their posts. Most managed to escape, but several were captured and later evacuated when the Chinese removed their own battle casualties.

Another attempt to regain the outpost was made by the Koreans at 1400, following artillery preparation and two air strikes. Three Marine attack squadrons, VMAs–323, –121, and –212 blasted the Chinese on the front slope of OP 36. The contour of the far side of the hill had provided the enemy a defiladed position and safety from 1st Marine Division organic weapons. But the MAG–12 air sorties, destroying many CCF automatic weapons and mortars and breaking up a company strongpoint, helped the Koreans counterattack and overrun the dazed defenders. Two KMC platoons, supported by artillery, mortar, and tank fire, then carried the OP after overcoming token Chinese resistance. After the enemy vacated OP 36, he still continued to remain in the low area to the northwest, close to the east side of the Sachon River. No serious attempt was made by the enemy to occupy the position for the rest of the month.

The 20-hour clash for control of OP 36 was believed to have developed from the Chinese ambition to occupy the position and thereby eliminate the harassing fires from Hill 36 that had struck CCF mainline troops. The 19–20 September attempts to wrest the outpost from Korean control resulted in an estimated 150 Chinese casualties, including 20 counted dead. KMC losses were placed at 16 killed, 47 wounded, and 6 missing.

On the day that the second September battle for OP 36 had ended, the Commandant of the Marine Corps had also just concluded his three-day visit and inspection of General Pollock's troops. Visiting every battalion in the division, General Shepherd was impressed by the morale and proficiency of the Marines, including the attached

1st KMC Regiment. During his visit to Korea, the Marine Corps Commandant was also presented the Order of Military Merit, Taiguk, by President Rhee. General Shepherd ended his Korean battlefront visit after a two-day inspection of 1st Marine Aircraft Wing units commanded by Major General Jerome (he had received his second star on 6 August).

More Enemy Assaults in Late September[17]

Even though the enemy had concentrated his strongest infantry attack in late September against the Korean Marines, his most frequent probes were launched against center regimental positions held by Colonel Layer's 1st Marines. Here the enemy was more consistent in conducting his defense. Chinese troops doggedly held on to the northern slopes of several Marine outposts, notably Hills 124 and 122. In this center regimental sector, the enemy initiated several attacks, the most significant of these occurring on the 20th.

This action against the left sector manned by 2/1 centered about Hill 124, where Lieutenant Colonel Batterton's battalion had established a 24-hour, squad-size outpost three days earlier. At 0345, Marines on Hill 124 observed two green flares fired from a hill about 1,100 yards to their front. At the same time the men of 2/1 observed numerous figures moving about downhill from their own position. It soon became evident that four enemy groups were converging on Hill 124 and preparing to assault the Marine defenses which shortly came under fire from enemy submachine guns and rifles. The main probe was a frontal assault against Batterton's men; it was made by about 20 Chinese and lasted only five minutes. Afterwards, all four assault groups withdrew but continued firing intermittently at the Marine squad. Nearly every Marine on the hill suffered wounds, most of these minor. Enemy losses for the action were placed at 22.

In this same sector Marines in late September attacked the northern slope of Hill 122, where the enemy still maintained a foothold. The proximity of Marine defenses at Bunker Hill to enemy positions, separated in some places by as little as 30 yards, was the cause of

---

[17] Unless otherwise noted, the material in this section is derived from: 1stMarDiv, 1stMar, 7thMar, 2/1 ComdDs, Sep 52.

frequent contact and clashes. Marines raided the enemy side of Bunker, using demolitions and portable flamethrowers to destroy trenches and bunkers, and their occupants. Tanks and artillery assisted in these brief offensive actions, usually undertaken at night. Flares were used frequently to aid in identifying and striking targets and in assessing the results.

It became routine during the last days of September for the Chinese to probe the Marine defenses at the Hills 124–122 axis. There did not appear to be a serious or determined assault to take either outpost, however. The Marines considered the infantry probes as just another form of harassment, although perhaps more personal and direct than the Chinese shelling, which inflicted daily losses. On the division right, Colonel Moore's 7th Marines, which had moved into this sector on 7 September, found enemy activities about the same. Artillery rounds caused the greatest number of casualties, although these attacks were not particularly spirited. Many enemy contacts occurred during the Marine combat patrols that largely characterized front-line operations at the end of September.

Chinese Intensify Their Outpost Attacks[18]

With the beginning of October, the 1st Marine Division became aware of certain changes that were occurring to its front. In the center sector, for the first time in two weeks there was no significant enemy ground activity, yet across the entire Marine front there was a build-up of enemy shelling. Part of the increased bombardment was directed at Hill 86 in the KMC sector, one of the positions recently cited as integral to the defense line in this area. Beginning at 2000 on 1 October, the Chinese broadcast a warning that they would knock down the outpost bunkers there unless the Korean Marines surrendered. When the KMCs manning the position did not, of course, surrender in reaction to this blatant propaganda tactic, the Chinese began showering Hill 86 with artillery rounds. During the next 20 hours, 145 rounds fell on and around the outpost. This

[18] Unless otherwise noted, the material in this section is derived from: FMFPac, *1stMarDiv Sum, Jul–Oct 52;* 1stMarDiv ComdD, Oct 52; 1stMarDiv G–3 Jnls, 1–7 Oct 52; 1stMarDiv PIRS 706–713, dtd 1–8 Oct 52; 1stMar, 5thMar, 7thMar, 11thMar, 3/1, 1/7, 2/7, 3/7 ComdDs, Oct 52; KMC Regt UnitRpts 214–220, dtd 2–8 Oct. 52.

incident marked the first time that the Chinese mainline forces had carried out an announced threat.

This type of operational tactic—first to warn, then to carry out the threat—was not, however, the reason for the increased Chinese shelling. Rather, as it turned out, the enemy was about to embark on a series of limited objective attacks against the division flanks, starting first with major outposts guarding the most critical terrain on the MLR. The artillery and mortar fire of the 1st had been but an initial step. At 1830 on 2 October, Communist direct fire weapons opened up from an area 2,800 yards northwest of OP 36, lashing all the KMC outposts within range. A tank platoon, dispatched to counter the fire, returned at 1915 without having located the hostile emplacements. Shortly after the tanks returned, an extremely heavy artillery barrage again fell upon all of the KMC regimental outposts. Ten minutes later, seemingly on the signal of one red and one green flare, the enemy guns lifted their preparatory fires to permit an infantry attack. The ground action simultaneously struck OPs 37, 36, and 86, the forward positions closest to the Sachon River.

At OP 37, the defending Korean Marine platoon fought valiantly for more than an hour against the assault of two enemy platoons, each of which required a company-size reinforcement before the Korean Marines were finally ousted. Although temporarily dislodged, they reorganized at the base of the position for a counterattack. Two counterattacks were made the next day, the second one carrying the Koreans to the top of the hill. Fierce enemy mortar and artillery shelling forced them to seek the shelter of the reverse slope before again renewing their assault. On 4–5 October, the outpost changed hands four times. At 1340 on the latter date, a heavy enemy artillery and ground attack compelled the KMCs to abandon their ravaged outpost; this withdrawal ended friendly control of OP 37 for the rest of the month.

Nearby OP 36 was also lost. In the course of the night the Korean Marines on OP 36 turned back two Communist assaults, but fell under the weight of the third. By sunup on 3 October, the exhausted Korean Marines were forced to give ground; the Chinese immediately occupied OP 36 and held it.

One more KMC outpost was to fall during the first week. OP 86 guarded the southwestern two-thirds of the regimental sector and freuently was the target of artillery shelling and ground attacks.

This position was also the most distant from the main line and the closest to the Sachon River.

The heaviest Communist attack on 2 October was against the KMC platoons defending Hill 86. Nearly a battalion of Chinese took part in this action, finally overpowering the outpost just before midnight. The defenders withdrew south to the bottom of the hill, where they were comparatively safe from enemy fire. Resting, receiving reinforcements, and regrouping during the early morning hours of the 3d, the Korean Marine force observed friendly artillery and air pound the outpost preparatory to their counterattack. It was made at 1015 and succeeded, after two hours fighting, in routing the Chinese from the outpost.

While the enemy was counteracting the ground loss with artillery and mortars, Marine air flushed out the Chinese, who had retreated only a short distance from the outpost. From atop the hill, Korean Marines witnessed many of the enemy hurriedly leaving the area under attack. This scattering of the enemy force prevented the Chinese from launching an immediate counterattack for control of OP 86 and gave the Korean Marines additional time in which to prepare their defenses. At 2200 on 6 October, an enemy force of undetermined size assaulted the position and wrested it from the Koreans before the end of the day. Early the next morning a KMC counterattack was successful, but at 0640 the Koreans were again compelled to withdraw, due to devastating blows from Chinese artillery. Loss of the third key outpost during the first week of October, ended for a time the flare-up of outpost fighting in the left regimental sector of the division front.

The middle part of the MLR, held in early October by the 1st Marines, received the least enemy attention in this period. Although frequent contacts were made with the enemy during the first part of the month, no outposts were lost. Most of the action was minor, *i.e.,* patrol engagements and Communist probes centered around Bunker Hill and Hill 124. Late on 5 October, a combat patrol from H/3/1 became involved in the most important ground action in Colonel Layer's area during early October. These Marines were surprised by a larger Chinese force lying in wait. The ambushers held their fire until the Marine combat patrol had cleared a small hilltop. At 2230, after a 20-minute fire fight, the patrol withdrew to the reverse slope of the rise, called in 81mm mortar fire, then broke contact, and returned to the MLR. There were 4 Marine casualties, and by count, 13 dead Chinese.

By far the greatest number of personal losses at this time occurred in the right area held by the 7th Marines, where the Chinese began a series of limited objective attacks against outposts guarding the division right flank. These offensives to obtain critical terrain in this sector, and others manned by the 1st Marine Division, would continue intermittently right up to the brink of the cease-fire, in July 1953.

In early October, Colonel Moore's troops manned nine permanent combat outposts. (See Map 14.) Seven of these had been taken over when the regiment relieved the 5th Marines in September. Two additional ones—Frisco and Verdun—had been established by the 7th Marines on the 14th and 26th, respectively. Of these nine forward positions, the Communists chose to concentrate on four, which formed a diagonal line roughly paralleling the center sector of the MLR at an average distance of about 450 yards. This quartet—Detroit, Frisco, Seattle, and Warsaw—together with Verdun,[19] at the 1st Commonwealth boundary, comprised the easternmost permanent outposts of the division. The first four positions were, on the average, slightly lower in elevation than the COPs in the regimental area to the west.

The frontline contest began with little forewarning other than a slight increase in enemy artillery and machine gun fire against Frisco and a light probe against Detroit. At 1836 on 2 October, the Communists launched a heavy artillery and mortar barrage against Seattle and Warsaw, and that part of the MLR nearest Seattle. Exactly one hour later, the preparation on the outposts lifted, permitting the enemy attack force to strike. Not less than a company assaulted the reinforced platoon on Warsaw, while a squad moved against the Seattle defenders. Warsaw fell in about 45 minutes,[20] Seattle held out five minutes longer.

---

[19] The outpost at the extreme right flank was given the name "Verdun" because of its World War I connotation of "They shall not pass." Col. Leo J. Dulacki ltr to Hd, HistBr, G-3 Div, HQMC, dtd 2 Jun 67, hereafter *Dulacki ltr.*

[20] During the latter stage of the fight for Warsaw, a Chinese soldier tossed a grenade into a bunker shared by five Marines. Private Jack W. Kelso, of I/3/7, quickly picked up the missile and ran outside with it. As he was throwing the grenade back to the Chinese, it went off in his hand. Disregarding his wounds, the Marine moved back inside the shelter, directed the other four to return to the MLR. and went oustide to cover their exit. As he was firing at the advancing Chinese soldiers, Private Kelso was hit several times by enemy bullets. His "conspicuous gallantry and intrepidity at the risk of his life" was later recognized in the posthumous award of the Medal of Honor.

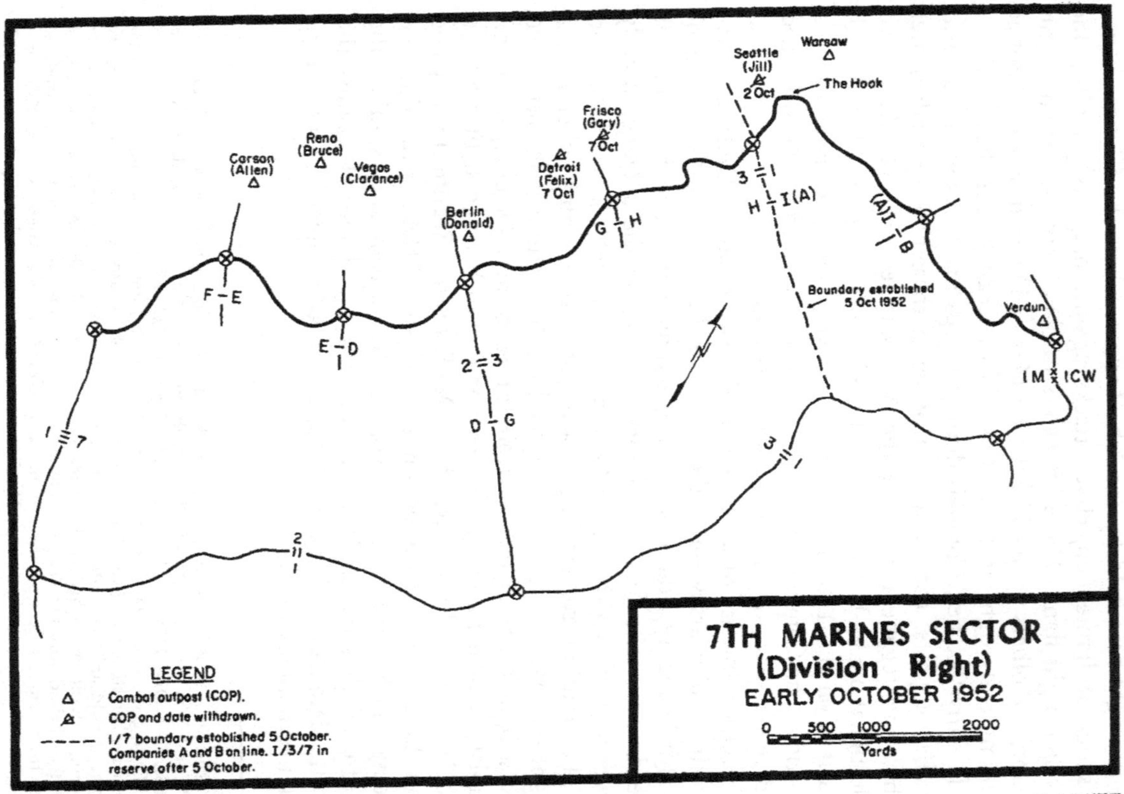

MAP 14

K WHITE

Immediately, plans for the recapture of both were made. At 2047, Captain John H. Thomas dispatched a platoon from his company, I/3/7, from the MLR to counterattack Warsaw. The platoon quickly took the position, for the enemy had withdrawn. At Seattle, the result was different. On 3 October, two squads from Company I departed JAMESTOWN at 0340, but came under enemy artillery fire en route to the objective. The squads worked their way forward nevertheless, but were unable to take the outpost. Captain Thomas then recalled the force, which reached JAMESTOWN at dawn. Later that day, just before dusk, air and artillery placed a smoke screen on Seattle while two squads advanced toward the outpost. When the counterattack met stiff resistance, a squad-size reinforcement[21] was sent from the MLR. Together the three units attempted to retake the position, but were forced to pull back because of heavy casualties. As the infantry again regrouped, Lieutenant Colonel Bert Davis, Jr.'s 2/11 fired preparatory barrages on the Chinese occupying Seattle. At 2225 the Marines assaulted the outpost again; as before, overpowering Chinese artillery and grenades inflicted such high casualties that the counterattackers were compelled to withdraw.

By this time, action at the two outposts had resulted in 101 Marine casualties, including 13 killed. By sundown on 3 October, the regiment had been forced off the two COPs and had been able to retake only one of them. Against Warsaw, the one that the Marines had recaptured, the Chinese immediately launched a counterattack. At 0145 on 4 October a platoon struck the position. This time the Warsaw garrison held, inflicting losses on the CCF and receiving none. The Chinese made an unsuccessful attempt against Frisco at 2300 on 5 October, when a squad attempted to drive the Marines from the outpost.

The enemy's repeated attacks and apparent determination to seize commanding terrain, plus the heavy casualties suffered by 3/7, led the 7th Marines to reinforce its MLR at 1200 on 5 October. At this time the right battalion sector then held by 3/7, was split into two sectors and the regimental reserve, 1/7 (Lieutenant Colonel Leo J. Dulacki) took over the far right of the 3/7 line, assuming respon-

---

[21] This squad was from Company A (Captain Frederick C. McLaughlin), which came under the operational control of 3/7 at 1130 on 3 October, relieving Company C (Captain Paul B. Byrum). The latter company had reported to the 3d Battalion from regimental reserve at 2130 the previous day. Company D was sent immediately to reinforce the hard-pressed Company I.

sibility for Warsaw and Verdun.[22] The 7th Marines thus had all three of its battalions on line with the regimental front manned, from the left, by 2/7, 3/7, and 1/7.

During the next 30 hours, the Communists launched a series of strong probing actions against the regimental outposts of the 7th Marines. Although the numerical strength used in these widespread limited objective attacks did not exceed that employed in previous large-scale outpost offensives, the scope of the operation on 6 and 7 October and the well-coordinated attacks indicated careful and detailed planning. Each move against the five outposts and two MLR positions attacked was preceded by unusually close attention to artillery and mortar preparation. This was to a degree unprecedented even when measured against those massive concentrations that had characterized Communist operations since the Chinese intervention in the war late in 1950.

Prior to the Communist general attack, the Marines made another attempt to retake Seattle. Leaving JAMESTOWN at 0600 on 6 October, a C/1/7 reinforced platoon was halted by solid resistance in the form of exploding artillery and mortar rounds. The forces returned to the MLR, reorganized, and jumped off again. At 0815, a two-squad reinforcement was dispatched from the main line. Meanwhile, the enemy, estimated at platoon reinforced strength, doubled his garrison, using troops from his outpost line. By 0900, a heavy fire fight was in progress, supported by artillery and mortars on both sides. Marines called on air in support of the attack, but the combined air and infantry action was unable to penetrate enemy defenses. Finally, at 1100, after five hours of close heavy fighting, the Marines broke contact and retired, bringing with them 12 dead and 44 wounded. Estimates of enemy losses totaled 71.

That evening, at dusk, artillery and mortar fire began falling on outpost positions across the entire regimental front and at two locations on the MLR. At the same time an estimated Chinese reinforced battalion in a coordinated effort advanced toward the Marine line and at 1930 assaulted the seven positions that had just been under artillery preparation. By midnight an estimated 4,300 rounds of artillery fire and 104 rounds of counterbattery fire had fallen on Marine positions. In the regimental left manned by 2/7 (Lieutenant

---

[22] At the same time one company, I/3/7, became the regimental reserve, having been relieved on the MLR at 1500 the previous day by A/1/7.

Colonel Caputo) the attacks appeared to be more of a diversion—merely probes by small units, which showed little inclination to press the attack. Carson, the most western COP held by the regiment, reported that the enemy soldiers withdrew at 2050. Two hours later Reno, the next outpost to the east, radioed to the MLR that the Chinese had just ceased their attacks at that forward post. A total of 12 Marines were wounded in these two actions.

On the far right, in Lieutenant Colonel Dulacki's sector, a reinforced CCF platoon poured over the Warsaw defenses at 1930. Immediately the outpost Marines called for the friendly artillery box. As these protective fires were being delivered all communication at the outpost was severed by hostile fire. Enemy artillery continued at a heavy rate. By 2000, however, communication was reestablished between the COP and MLR. The first message from the besieged outpost was a request for more artillery. With additional fire support and continued stiff outpost resistance, the Chinese at 2055 relinquished their quest to regain Warsaw.

The enemy's most determined assaults on the night of 6–7 October were made upon a pair of outposts, Detroit and Frisco, manned by the middle battalion, 3/7 (Lieutenant Colonel Gerald F. Russell). Two JAMESTOWN areas in this sector were also attacked, but only briefly. The assault against the outposts was executed by a Chinese battalion which sent one company against Detroit and another against Frisco, east of Detroit. Both outposts were manned by two squads of Marines.

At Detroit, the Company G Marines reported that the initial attack made at 1940 on 6 October by a Chinese company had been rebuffed. The enemy did succeed, however, in advancing to the outpost trenchline. Strong defensive fires prevented him from exploiting this initial gain by occupying any of the bunkers, and the attackers were forced to pull back. After regrouping, the Chinese returned at 2100 and again were able to secure a foothold at the main trench.

Marine artillery assisted the outpost defenders in repulsing this new attack, but not before Chinese interdictory fires had disrupted all communications between the COP and its MLR support company. Some Chinese had also moved south in the vicinity of the MLR, but these attacks were neither persistent nor heavily supported. At 2115 the last of the enemy intruders had withdrawn from the MLR. At about this same time, 3/7 heard Detroit request overhead VT fires, but shortly after this the battalion again lost contact with the outpost.

Two squads were then sent out to reinforce the position. They were stopped, however, by heavy Chinese artillery barrages. At the outpost, Marine artillery fires had forced the Chinese to retreat, but at 0015 the enemy reappeared at the trenchline. The artillery regiment once again applied the overhead fire remedy, but with less success—the Chinese, neither retreating nor advancing, took cover in the trenches. During the long night, attempts to reestablish communications with Detroit had proved fruitless, although battalion radio operators reported that they had heard Chinese language coming over one of the Marine radio nets used by the COP. A six-man reconnaissance detail was sent forward to investigate. It returned at 0355 with the information that Detroit was now held by the enemy. Two wounded Marines had escaped; the rest of the Detroit garrison had fallen to the enemy. At 0630 the Marines withdrew after heavy fighting that had lasted more than 10 hours.

During the earlier part of the night, while the battle for outpost control raged at Detroit, reinforcements had also been dispatched to Frisco to help stabilize the situation at this adjacent Company H/3/7 outpost. Like Detroit, it had been attacked by a Chinese company, beginning about 2000. An hour and a half later some of the enemy had made their way into the trenchline, but were repulsed with the help of friendly artillery VT. Shortly after midnight the enemy again probed Frisco and reached the trenchline. At 0115, two squads jumped off from JAMESTOWN, but a rain of Chinese artillery interrupted their progress. Throughout the early morning hours of 7 October, Company H and I units were sent out from the MLR to buttress the Frisco defense and stem the enemy attack. At 0510, a reinforced platoon from the reserve company was sent to renew the counterattack. It was this Company I unit that finally restored control of the COP to the Marines.[23] Another reinforcing

[23] During the predawn attempt to retake Frisco on 7 October, Staff Sergeant Lewis G. Watkins, I/3/7, although already wounded, led his rifle platoon in the assault against Frisco. When an enemy machine gun impeded their progress, Staff Sergeant Watkins grabbed a wounded man's automatic rifle to help get the assault moving forward again. At that instant, an enemy grenade landed in the midst of the Marines. Staff Sergeant Watkins immediately seized it. Just as he was about to hurl it away it exploded in his hand. The grenade took the sergeant's life but he had saved his fellow Marines. For his bravery Staff Sergeant Watkins was posthumously awarded the Medal of Honor.

platoon arrived at the outpost just as the Marines there had evicted the remaining Chinese assault forces. At 0715, 7 October, Frisco was declared secure.

Its precarious position, however, demanded either an investment of more outpost troops to retain possession of it or else its abandonment, in conjunction with other measures to neutralize loss of the position. At 1804 that day the latter course was instituted. The 7th Marines reported that the enemy had suffered an estimated 200 KIA and unknown WIA as a result of the bitterly contested outpost attacks on 6–7 October. Marine casualties were listed as 10 killed, 22 missing, 105 wounded and evacuated, and 23 not-seriously wounded.

In all, during the first week in October, the 1st Marine Division gave up six outposts, or forward positions, that had been sited on some of the commanding ground in the Marine area. On the division left, COPs 37, 36, and 86 were the ones most removed from the Korean MLR and thus easily susceptible to being overrun by the enemy at will and to his early reinforcement.[24] The division theorized that near winter and the subsequent freezing of the Sachon would facilitate the movement of Chinese troops and supplies across the river to new positions. The enemy was now able to operate patrols east of the river without interference. At the opposite side of the division MLR, on its right flank, Detroit, Frisco, and Seattle had been lost. By gaining this string of outposts, the enemy was better able to exert pressure against other Marine positions forward of the line and the critical ground on JAMESTOWN.

To counter this threat, General Pollock strengthened the outposts close to the MLR and increased his patrolling requirements. It was decided that in some cases the mission of the COP—that of providing early warning of impending attack and slowing it down—could be accomplished as effectively by using patrols and listening posts at night.

By these activities, the Marines hoped to minimize the Chinese gains and prevent the launching of new attacks against either division COPs or JAMESTOWN. The serious situation on the outposts was compounded by existing political considerations, which prevented the Marines from initiating any real offensive campaigns. Moreover, any

---

[24] FMFPac, *1stMarDiv Sum, Jul–Oct 52.*

hill taken was invariably backed up by a still higher one, controlled by the enemy. The key factor was not so much holding an individual outpost as it was to insure that the enemy was unable to penetrate the JAMESTOWN line.

More PRESSURE, *More CAS, More Accomplishments*[25]

Some of the enemy ground pressure against the outposts in September and October had been relieved by the increase in the number of air strikes received by the 1st Marine Division. De-emphasis of the Air Force interdiction strategy in favor of striking the enemy wherever (and whenever) it hurt him most had made available more aircraft for close support of ground operations.[26] The UN commander, General Clark, who had given the green light to the shift in USAF policy and targets, followed the giant hydroelectric strike in June with a mass attack the next month on 30 military targets located near the North Korean Capital. During a year's freedom from air attack (July 1951–July 1952) Pyongyang had become not only the major logistics center for combat equipment and personnel but also the focal point for command and control of Communist ground and air defense efforts.

Designated Operation PRESSURE PUMP, the 11 July strike against Pyongyang called for three separate attacks during daylight and a fourth at night. This extended time over the target would give enemy fighters more than ample time to take to the skies in defense of the Capital. Because Pyongyang "was defended by 48 guns and more than 100 automatic weapons, making it one of the worst 'flak traps' in Korea,"[27] there was considerable hazard in the operation. Added

---

[25] Unless otherwise noted, the material in this section is derived from: *PacFlt EvalRpt* No. 5, Chap. 9; 1st MAW ComdDs, Jun–Oct 52; MAG–12 ComdDs, Jun, Sep 52; MAG–33 ComdD, Aug 52; MACG–2 ComdD, Sep 52; VMA–312 ComdDs, Sep–Oct 52; VMA–323 ComdDs, Jun–Jul, Sep 52; VMF(N)–513 ComdDs, Jun–Jul 52; VMJ–1 ComdD, Jul 52; Cagle and Manson, *Sea War, Korea;* Clark, *Danube to Yalu;* Field, *NavOps, Korea;* Futrell, *USAF, Korea;* Rees, *Korea.*

[26] The 1st MAW chief of staff during this period, then Colonel Samuel S. Jack, offered the opinion that "the Fifth Air Force was most sympathetic to Division requirements for close air support from Wing sources. The Eighth Army in the Joint Operations Center proved to be the principal limiting factor in the assignment of air in accordance with these requests. Also, requirements that Division CAS requests filter through I Corps and JOC constituted a major factor in Wing response." *Jack ltr.*

[27] Futrell, *USAF, Korea,* p. 482.

danger to the pilots resulted from the decision to forewarn the North Korean civilian population of the air assault. General Clark explained the reason for dropping warning leaflets prior to the attack on Pyongyang:

> The objective was in part humanitarian and in part practical. We had to hit Pyongyang because the Communists had made it a major military headquarters and stockpile area. We wanted to warn the people away from danger areas. By warning them away we disrupted their daily lives and made it difficult for the Communists to maintain any kind of schedules in their work in the city.[28]

Results indicated that both the destructive and the psychological aspects of the mission were successful. American, British, and ROK planes completely destroyed 3 of the 30 military targets attacked. Of the rest, only two escaped major damage:

> According to . . . reports, the North Korean Ministry of Industry's underground offices were destroyed and a direct hit on another shelter was said to have killed 400 to 500 Communist officials. Off the air for two days, Radio Pyongyang finally announced that the 'brutal' strikes had destroyed 1,500 buildings and had inflicted 7,000 casualties.[29]

Of the far-reaching effect of the leaflets, the UN commander later wrote:

> The warning leaflets, coupled with the bombing, hurt North Korean civilian morale badly. The very audacity of the United Nations in warning the Communists where bombers would strike hurt morale because it emphasized to the North Koreans just how complete was UN mastery of the air. Contrarily, it made them see even more clearly that the Communists were ineffectual in their efforts to ward off our air blows. . . .
>
> As a result of the warnings, the bombings, the failure of the Communists to provide protection, and the refusal of the Communists to permit evacuation of the clearly defined target areas, civilian resentment was channeled away from the UNC bombers and towards the Communist rulers.[30]

The record set by the 1,254 sorties flown in this 11 July operation was to last only seven weeks. On 29 August, 1,403 sorties were employed in a new strike against the Capital. The massed raids against military targets in Pyongyang, known as the "All United Nations Air

---

[28] Clark, *Danube to Yalu,* pp. 208–209.

[29] Futrell, *USAF, Korea,* p. 482.

[30] Clark, *Danube to Yalu,* p. 209. "I told you so" leaflets were dropped after the raid to impress the inhabitants with the importance of believing the warning leaflets. USAF, *Ops in Korea,* No. 127, pp. 36, 37.

Effort" turned out to be the largest one-day air assault during the entire three years of the Korean War. Again attacking at four-hour intervals three times during daylight, Allied aircraft blasted a list of targets that "read like a guide to public offices in Pyongyang and included such points of interest as the Ministry of Rail Transportation, the Munitions Bureau, Radio Pyongyang, plus many factories, warehouses, and troop billets."[31] Of the 45 military targets in the city, 31 received moderate-to-severe damage according to post-strike photographs.

Substitution of the previous interdiction strategy by PRESSURE attacks brought increased close air support to frontline troops. As a result of this expanded number of CAS sorties, wing pilots and ground forward air controllers greatly increased their operational proficiency.[32] The Marines were still not satisfied with the close support picture, however, and neither were a number of U.S. Army commanders. Some of the latter regarded General Clark as the champion of more extensive close air support missions for frontline units, but he quickly dispelled this view. Instead, he cautioned these supporters of Marine-type close air support to accept the existing procedures, which were derived from the "vast reservoir of experience . . . [representing] the composite view of senior members of the Armed Forces [with] the longest and most responsible experience in close support during World War II."[33] At the same time the UN commander, on 11 August 1952, had advised his force commanders to study the factors affecting the close air support situation in Korea and comment on certain UNC proposals for improving the CAS system.

In the close air support picture for the Marines, October was a bright month. In the outpost battles of early October, the 1st MAW put 319 sorties in the air during both day and night to strike, strafe, bomb, and burn enemy positions and troops facing General Pollock's division. A new level of achievement had been reached during the Bunker Hill battle in August. That month nearly 1,000 aircraft, predominantly Marine, loosed ordnance at targets on and near the Chinese MLR and OPLR.

During the first six months of Marine ground operations in defense of JAMESTOWN, wing squadrons and pilots had made major contri-

---

[31] Futrell, *USAF, Korea*, p. 489.
[32] *PacFlt EvalRpt* No. 5, p. 9–53.
[33] *Ibid.*, p. 9–143.

butions to the U.S. air effort in Korea. On 7 June 1952, First Lieutenant John W. Andre, VMF(N)–513, piloting a World War II model Corsair on a night armed reconnaissance mission over the west coast of North Korea, shot down an enemy piston-driven Yak fighter. It was the first time that a Russian-built plane of that model had been knocked out of the skies at night by another plane. This aircraft was also the fifth kill for the lieutenant, making him the first Marine nightfighter ace in Korea.[34]

Nearly three months after that record, another one emerged: the first Marine to down an enemy jet with a propeller-driven aircraft. Late on the afternoon of 10 September, Captain Jesse G. Folmar and First Lieutenant Willie L. Daniels, both of VMA–312, had taken off from the *Sicily* to attack an enemy troop concentration reported to be south of Chinnampo, on the west coast just below the 39th Parallel. Shortly after reaching the vicinity of the target, the Marine Corsairs were jumped by a pair of MIG–15s. Two more Russian-made jets tore into the fight. During a fast exchange of cannon and machine gun fire, the Marine captain was able to score lethal hits on one of the MIGs. When four more of them picked up the chase, the vastly outnumbered Marines broke for home, heading westward in a diving turn.

Captain Folmar's return to the *Sicily* was delayed almost immediately:

> I had just started picking up good diving speed when I saw balls of tracer ammo passing on my left and at the same instant felt a severe explosion in my left wing . . . I saw that the left aileron and four feet of my left wing were gone.[35]

This damage caused the plane to rapidly go out of control. While still able to maneuver, the Marine aviator headed for the sea and as he neared it, bailed out of his Corsair and parachuted into the ocean. A rescue plane out of Cho-do picked him up and returned the captain, who had sustained a slight shoulder injury, to the carrier. Lieutenant Daniels, who had alerted the rescue force, circled his descending flight leader until he hit the water. After ascertaining that the waterborne flier's condition was satisfactory, the lieutenant turned his plane towards the *Sicily*. In a short while he was safely home.

[34] The first Marine night ace was Captain Robert Baird, who shot down six Japanese planes between 9 June and 14 July 1945. Sherrod, *Marine Aviation*, p. 404. Lieutenant Andre's first four planes were also downed during World War II. See Appendix F for Marine air kills during the Korean War.

[35] VMA–312 ComdD, Sep 52.

In late September, Major Alexander J. Gillis, VMF–311, assigned earlier that summer to the Air Force's 335th Fighter-Interceptor Squadron, 4th Fighter Group, as an exchange pilot.[36] distinguished himself by becoming not only the first naval aviator to destroy three enemy aircraft in Korea but also the second one to get a multiple killing in a single day.[37] Flying in a four-plane Sabrejet formation near the vicinity of the mouth of the Yalu on 28 September, Major Gillis led another plane after two MIG–15s. By superior pilot technique and aggressive tactics, he forced one of the enemy to crash during a low altitude chase. Later on during the sortie, the Marine initiated an attack on a solo MIG, closing on it and scoring hits that caused the plane to become uncontrollable and the pilot to eject. Major Gillis also had to eject from his F–86 after it became disabled by the MIG. The incident had occurred on the Marine aviator's 50th combat mission with the Air Force. He spent nearly four hours in the Yellow Sea before a rescue helicopter picked him up.

Another feat, this one a study in determination and perseverance, had occurred early in the summer. On 22 July, the VMJ–1 commander, Lieutenant Colonel Vernon O. Ullman, had taken to the air for a photo mission over North Korea in the vicinity of Sinanju, located near the Yellow Sea 40 miles above Pyongyang. During the first of seven scheduled flights, he encountered heavy flak but nevertheless completed his first mapping run in the area. Further, the Marine flier decided that the antiaircraft menace was not going to force him to abandon the remaining part of his task. He continued. On the second of his seven runs, some 40 enemy jets (MIG–15s) appeared on the scene. These were dissuaded from close-in interference, however, by the photo escort of 24 USAF single-engine Sabrejet fighters. Thereafter, the Russian-made aircraft disappeared; Lieutenant Colonel Ullman continued, despite the intense, accurate enemy antiaircraft fire, until he concluded his mission.

---

[36] The exchange program "appears to have originated with the participation—at Tactical Air Command's invitation—of two Marine Corps and two Navy pilots . . . in the fall of 1947." Within two years, the program designed to "indoctrinate selected Air Force and Navy pilots in the air operational and air training activities of each other's service, had received Department of Defense approval." On 1 October 1949 the program went into effect. Initially the exchange period was one year, but after the Korean fighting broke out, the period was reduced to approximately three months. Marine participation began late in 1951. Atch 1 to Hq, USAF (AFCHO) memo to Maj J. M. Yingling, HQMC, dtd 16 Jan 67 in v. V, Korean comment file.

[37] On 15 September, Major Gillis had shot down a solo MIG–15.

The type of determination displayed by Lieutenant Colonel Ullman helped Marine tactical squadrons achieve some kind of distinction nearly every month from late spring to the fall of 1952. In May, VMF–323 ("the Death Rattlers"), then commanded by Major William A. Weir, established a squadron one-month record for number of combat sorties, 1,160, and total combat hours, 2,362.7. A high percentage of aircraft availability, 95.6, helped make this mark possible. On 1 June, VMA–312 received the congratulations of CTF 95 for its "outstanding performance under difficult conditions" during the spring months. During this period the squadron, based on board the USS *Bataan,* had been particularly hampered by excessive turn-over of key squadron officers and flight leaders. This continual squadron rotation resulted in considerable variation in pilot indoctrination and need for field carrier landing qualification, due to the "close tolerances in pilot skill required by carrier operations."[38] Despite these difficulties, VMA–312 had scored an impressive 80-sortie mission, flown by 24 aircraft, on 18 April.

Additional recognition of professional excellence was conferred upon Marine squadrons in July. On the 17th, the senior advisor to the ROK I Corps expressed the gratitude of the corps commander for the magnificent support the 1st MAW pilots had provided during the second week of the month. All four attack squadrons in MAG–12 and both fighter units in MAG–33 had taken part in these CAS missions. A week later, eight planes from Lieutenant Colonel Henry S. Miller's VMA–323, (which, along with Lieutenant Colonel Graham H. Benson's VMA–212, had been redesignated from fighter to attack squadrons the previous month), completed an unusually successful interdiction mission at Hago.

Located 25 miles northwest of Kaesong, the village reportedly was the site of heavy troop concentrations, active mortar positions, and antitank weapons. Leaving K–6 at 1725, the eight Marine VMA–323 pilots were soon over the target. Comprising the Death Rattler's flight were Majors John M. Dufford, Raymond C. Holben, William H. Irvin, Jr., and Curtis E. Knudson; Captain John Church, Jr.; First Lieutenant William A. Poe, Jr.; and Second Lieutenants Stuart L. Spurlock and James S. Thompson. At 1810 their attacks were launched, using 1,000-pound bombs, napalm, rockets, and 20mm am-

---

[38] *PacFlt EvalRpt,* No. 4, Chap. 10, p. 10–77.

munition. The strike was over almost as soon as it had started, and when the Marines departed, not one building remained in useful condition. But it was not until several days later that the final results of the strike were known. Intelligence sources reported that the raid had caught the enemy troops at the evening meal; more than 500 had been killed by the Corsairs, aptly called "Whistling Death" by the Japanese in World War II.

For the remainder of the summer and into the fall Marine groups and squadrons continued their record-breaking and efficient support of ground troops and naval forces. With four squadrons (two day, one night-fighter, and one photo), MAG–33 sent 141 sorties against the enemy on 6 August. This one-day group record occurred just before the departure of Colonel Condon, who turned over the reins of the organization to Colonel Herbert H. Williamson on the 11th, and then took command of MAG–12.

Shortly before Colonel Condon relinquished command, he was particularly pleased by the success of a four-plane strike by VMF–311 (Major William J. Sims) in support of the U.S. 25th Infantry Division commanded by Brigadier General Samuel T. Williams. Major Johnnie C. Vance, Jr., strike leader, was accompanied in this flight by Captain George R. Brier and Second Lieutenants Charles E. Pangburn and Whitlock N. Sharpe. Up until this time the infantry had been particularly harassed by several enemy frontline fortifications and supporting artillery. The four pilots destroyed three bunkers and two heavy guns and also caved in approximately 50 feet of trenchline on the 7 August strike. Upon learning of the success of the Marine pilots and the conditions under which the attacks were carried out—dangerous terrain and constant ground fire directed towards the planes—the general dispatched a letter, commending the "skill, courage, and determination displayed by these pilots. . . ."[30]

Another congratulatory message was received in September, this one from General Pollock for the excellent support given by MAG–12 on the 20th. With three attack squadrons participating, Colonel Condon's group had neutralized Chinese weapons and troops at OP 36 to help prevent a takeover of the Korean position. The pilots reported well over 100 Chinese casualties. Lieutenant Colonel Kenneth R. Chamberlain's VMA–323 contributed most of the 23 Marine

---

[30] MAG–33 ComdD, Aug 52, p. 16.

sorties. The other attacking squadrons were VMA–121 (Lieutenant Colonel Wayne M. Cargill, who 10 days earlier had relieved Lieutenant Colonel Crawford), and VMA–212, commanded by Lieutenant Colonel Maurice W. Fletcher.

September was a month of mixed fortunes in the air war over Korea. The successful CAS strikes of the 20th followed only a few days after another high point set on 14 September, when Lieutenant Colonel Cargill's attack squadron flew its 5,000th combat sortie since arrival in the Korean theater in October 1951. Then on 15 September, General Jerome commissioned a new kind of unit in the wing, Marine Composite Squadron 1 (VMC–1), whose mission was to provide electronic counter-measures (ECM) for UN aircraft. Lieutenant Colonel Lawrence F. Fox headed the squadron, the only one in the naval service with an ECM primary mission in Korea.

Three days after the commissioning, a strange incident transpired. North of the UN line and at an altitude of 9,500 feet, a F–84 Thunderjet fighter, with U.S. Air Force markings and insignia, attacked a propeller-driven Air Force trainer. The slower plane immediately began defensive maneuvering, flying in tight circles. After making five turns, the trainer pilot saw the supposedly friendly jet fly off.

It was believed that such a paradoxical occurrence was due to the substantial number of F–84 losses and the enemy's ability to piece together and fly an aircraft of that model. A few similar episodes—attacks by apparently friendly aircraft on UN planes—had previously taken place. In each case, the impostor was a model of U.S. aircraft that had suffered particularly heavy losses.

Another incident in September had dire consequences. On the 10th, MAG–33 dispatched 22 fighter aircraft from VMF–115 (Lieutenant Colonel Royce W. Coln) to attack reported troop concentrations near Sariwon, 35 miles directly south of Pyongyang. The F9F Panther jets had completed the strike and were returning to their K–3 base when they were diverted to land at K–2, Taegu, where the weather was better. Fog had suddenly swept over the field at K–3, reducing visibility to zero. Sixteen Panthers landed safely at K–2, 45 miles southwest of the Marine field at Pohang. The remaining six, piloted by Majors Raymond E. Demers and Donald F. Givens, First Lieutenant Alvin R. Bourgeois, and by Second Lieutenants John W. Hill, Jr., Carl R. Lafleur, and Richard L. Roth, flying in formation in poor weather, crashed into the side of a 3,000-foot mountain while descending.[40] They would have required only an additional 600 feet of altitude to clear the summit.

Losses of Marine pilots and aircraft had been of growing concern to the wing command. The initial success of the flak suppression fires had eliminated the one successful Communist source of air defense, accurate antiaircraft firing. One result was that noncombat accidents for a while during the summer became the principal cause of pilot and plane attrition. To help reduce these operational accidents as well as the combat losses, the two Marine air groups instituted squadron training programs and also directed the adoption of several new corrective procedures. In MAG–12, for example, a study of results from the FAF policy that limited bombing runs to one for interdiction and two for CAS targets revealed a sharp reduction in hits from flak. Tactical squadron commanders in MAG–12 drew up a syllabus during September to test proposed defensive tactics for their propeller aircraft to employ against enemy jets. The carrier squadron, VMA–312, began that same month the additional practice of field carrier landing qualification at K–6 for new pilots before permitting them to operate from the carriers.

In spite of these efforts, pilot losses spiralled alarmingly in October. For the rest of 1952, the monthly totals remained near that month's level. On the other hand, aircraft losses during October dropped sharply to 10 from the September total of 22. This lower figure was not to be exceeded until May 1953. These remedial procedures were considered at least partially responsible for the substantial decrease in aircraft losses.

In another area, a mid-October landing at Kojo, on the east coast immediately south of the 39th Parallel, did not work out as planned. The amphibious operation was in reality a feint intended to draw troops away from frontline positions and expose them to naval air and gunfire as they rushed in reinforcements. The enemy failed to rise to the bait, and actually only a few Communist troops were sighted. VMA–312 provided armed reconnaissance, tactical air operation, and naval gunfire spotting during the feint. Although they made little enemy contact, the Marine "Checkerboard" pilots operat-

---

[40]Although not definitely proven, there were "some indications of false radio beacons being used by the enemy in clandestine operations in the K–2 area." *Jack ltr.*

ing off the *Sicily* gained much experience in landings and take-offs under the adverse conditions of rough seas and high winds.

Rockets, Resupply, and Radios[41]

Through October 1952, operational control of Korean based Marine fighter and attack squadrons was still vested in commanders other than General Jerome. Tactical squadrons continued to be directed by the FAF or Navy in their missions; the observation and helicopter squadrons were under operational control of the 1st Marine Division and utilized, as before, at its discretion.

HMR–161, commanded since 8 August by Lieutenant Colonel John F. Carey, continued its primary mission of evaluating rotary wing aircraft and their methods of employment. One tactical innovation, movement of elements of the 4.5-inch Rocket Battery, was undertaken during August soon after the Bunker Hill battle. With ground-fired rockets, the problem of a tell-tale cloud of dust and brilliant flash of the rockets after each salvo had always plagued the artillerymen. This seldom went unnoticed by the enemy, who often showered the marked area with counterbattery fire. On 19 and 20 August, in Operation RIPPLE, HMR–161 and the rocket battery proved that these two units could successfully shoot and scoot to a new location and fire effectively again without drawing an enemy reprisal. This Marine Corps innovation in air mobility—the first displacement of field artillery under combat conditions—offered a major time-saving advantage. Whereas previously it took approximately a half-hour for rocket launchers to move from their bivouac area to firing position,[42] deployment by helicopter could be made in a matter of minutes, a time factor that could be critical in event of an enemy attempted breakthrough.

The operation demonstrated that helicopters not only could transport rocket crews with weapons and ammunition to firing areas far more rapidly than conventional wheeled vehicles, but that the rotary craft could airlift these weapons into places inaccessible by road. The nature of the mountainous terrain proved advantageous in that

---

[41] Unless otherwise noted, the material in this section is derived from: *PacFlt EvalRpt* No. 5, Chaps. 8, 9; 1st MAW ComdD, Oct 52; HMR–161 ComdDs Aug–Sep 52.
[42] *Henderson ltr III.*

hills and valleys provided defiladed areas for loading and firing the weapons as well as protected routes for helicopter movements. An observation made by pilots for operations in other types of environment, not offering as much cover and concealment, was that the shiny blue paint on their birds would make detection easy in most surroundings and that camouflage paint would lessen the risk from enemy AA.

Transport helicopters of HMR–161 continued to augment those of VMO–6 in casualty evacuation and ferrying Marines and other front-line troops. The observation squadron maintained its policy and outstanding record of emergency flights of the wounded under any weather conditions except dense fog (electronic navigational aids still were not available). In August, various mechanical failures developed among the newly received HO5S–1 Sikorsky helicopters. These three-place observation aircraft were underpowered but superior in many flight characteristics to the HTL–4 helicopters then in the squadron. Mechanical difficulties with the newer aircraft increased until it became necessary to ground them late in October until replacement parts became available in the supply system.

Employment of transport helicopters for logistical support contined to be a principal use of such rotary wing aircraft as the end of 1952 approached. Tests earlier in the year had proved the theory that this versatile aircraft could resupply a battalion manning the MLR. The next step was to determine if the logistical support for an entire combat regiment could be accomplished by helicopter. Operation HAYLIFT, conducted during 22–26 September, the last of five operations that month for HMR–161, was to test and evaluate helicopter resupply of Colonel Moore's 7th Marines. Plans called for delivering all Class I, III, and V items and such Class II and IV items as could be accommodated. Two loading and four unloading sites were prescribed. All but extremely valuable cargo, such as mail, was to be carried externally in slings or wire baskets.

HAYLIFT did show that at least for a short period of time—five days—a helicopter squadron, utilizing 40 percent of its aircraft, could sustain a MLR regiment. Following the general procedures employed previously with the battalion, HMR–161 found that no great changes were necessary for resupply of the regiment. Two recommendations emerged from an evaluation of HAYLIFT. One stressed the need for establishment of an operations center manned by representatives of each unit participating in the exercise. The second called for development of a more flexible loading system, one that would permit rapid weight increases or decreases of 50 pound increments, as the situation demanded. Such a method would make possible a more efficient payload for each lift.[43]

Transport on a larger scale in the 1st MAW was accomplished by General Jerome's few transport aircraft reinforced by the eight R5Ds from the VMR–152 detachment. In June, the passenger-carrying operations reached the peak for the entire Korean War; that month, 17,490 troops and military-associated civilians utilized the reinforced wing transport aircraft. June 1952 was also the second busiest month in freight transportation (7,397,824 pounds, nearly double the figure for June 1951).

Squadrons that were unable to better their performance records in some cases could trace their trouble to the inability to get all of their planes off the ground. Several models were subject to spare parts shortages.[44] New aircraft, the F3D–2s and the AU–1s received in June by VMF(N)–513 and VMA–212, respectively, had preceded an adequate stocking of normal replacements for worn out or defective parts. The night fighter squadron was handicapped also by introduction into the supply system of inadequate radio tubes, which burned out rapidly. The most critical shortage, however, was parts for starter units of jet engines. This deficiency was not corrected until summer. One problem never quite eliminated was the confusion of supply orders intended for the helicopters in HMR–161 and VMO–6. It was believed that the close resemblance of Sikorsky HRS and HO5S part numbers and nomenclatures had caused the improperly-marked requisitions and mix-up.

The 1st Marine Division logistical situation during the summer and fall of 1952 was generally excellent. General Pollock's units did not suffer from the shortage of spare parts experienced by the 1st MAW whose aircraft sometimes had to be grounded because of a missing spare part. U.S. Army support in the replacement of

---

[43] For example, on 25 September, rain soaked the cardboard cover of the rations, adding extra weight to each preloaded lift of these Class I supplies. On the other hand, a heavier load could have been used at times. As the helicopter used up its fuel, a commensurate increase in cargo could have been carried.

[44] Spare parts shortages are "inherent in the introduction of new equipment into the field and prior to the development of usage data." a major effort was made at this time by 1st MAW to improve its critical spare parts support by improved stock control procedures and complete inventory. *Jack ltr.*

worn-out Marine vehicles for new Army ones proved satisfactory. No major problems arose in engineer support. Medical evacuation and treatment and the level of supplies in the five companies of the 1st Medical Battalion remained excellent.

There were two significant changes in the logistical support provided the Marine division early in the fall. One dealt with employment of the division's 1st and 7th Motor Transport Battalions, located in the rear support areas. Beginning in September, the companies were placed in direct support of the four infantry regiments, with liaison by Lieutenant Colonel Kenneth E. Martin, division motor transport officer. It was believed this decentralization would have the following advantages:

1. Decreased vehicle mileage and therefore less driver fatigue and prolonged vehicle life.
2. Increased dispersal as a safeguard against loss of wheeled vehicle support in event of an unexpected and successful enemy attack.

The other change was a shift in the emphasis of support rendered by the Korean Service Corps. During October, each of the three frontline regiments received 300 more laborers, raising the total to 800. Rear area units paid for the increase, since the KSCs were detached from support units and sent forward to the MLR.

Logistical support from the 1st Signal Battalion left little to be desired. Commanded by Lieutenant Colonel John E. Morris[45] when the Marines moved to western Korea, the signalmen helped establish and maintain an extensive communications net, with 5,200 miles of wire within the division and several vital links to adjacent and higher commands.[46] Wiremen worked around the clock to lay and maintain the telephone lines, which suffered considerable damage from the artillery and mortar barrages. When possible, the signalmen raised the wires off the ground. The battalion set in more than 1,400 telephone poles. After the system had been installed and was working efficiently, the July floods washed away part of the major communi-

---

[45] On 4 April Lieutenant Colonel Alton L. Hicks assumed command of the battalion; Lieutenant Colonel Jacob E. Glick relieved him on 3 August.

[46] Communication with General Kendall's I Corps consisted of radio-teletype, telephone, radio relay, courier plane, and motor messenger. *PacFlt EvalRpt* No. 5, p. 8–68. The 11th Marines also had an additional 1,100 miles of communication wire. *Henderson ltr III.*

cations. By improvising and by setting up emergency equipment, the battalion was able to maintain the flow of communications traffic at a satisfactory level. Replacement items were provided by the U.S. Army on a reimbursable basis in accordance with existing directives.

In September it became apparent that the signal equipment used to maintain division communications was no longer equal to the demands placed upon it. The extensive ground area plus the number and size of reinforcing units had not only put a heavy burden on radio, telephone, and teletype equipment but also caused the depletion of reserve stocks. With the spare equipment in use, there was no pool to draw upon when units turned in defective equipment for repair. Neither were there available replacements for materiel destroyed by enemy action. Items most urgently needed were flown in from the States. Other critical parts came from Army sources in Japan and Korea. By the end of October, the communication resupply had returned to a more normal condition.

Before the month ended a different type of critical situation was to confront the division. It appeared that the enemy's success in seizing a half-dozen outposts earlier in October had only whetted his appetite for more. Chinese eyes were turned towards positions that held still more potential value than the stepping-stones just acquired. The extreme right battalion in the division front held by the 7th Marines was the focal point of the new effort.

CHAPTER V

The Hook

Before the Battle—Preparations for Attack and Defense—Attack on the Hook—Reno Demonstration—Counterattack—Overview

Before the Battle[1]

AFTER THE HEAVY FIGHTING in early October, there was a change in the 1st Marine Division dispositions. On the 12th, the 5th Marines relieved the 1st in the center sector and the latter regiment went into reserve. For the next two weeks the lull that prevailed across the regimental front was in sharp contrast to the intense fighting there earlier in the month. On the division left, the Korean Marines, not engaged in any sizable Communist action, conducted frequent tank-infantry reconnaissance patrols and ambushes forward of their MLR. In the center of the division line the 5th Marines, too, found their Chinese opponent seemingly reluctant to pursue any combat offensives, though his harassment of the Bunker Hill area represented the strongest action against the Marine division at this time. The 7th Marines, holding down the right sector, similarly encountered the enemy for only brief periods, these contacts during patrol actions lasting no more than 15 to 30 minutes.

Upon its relief from the MLR, the 1st Marines took over the division rear area. There the regiment continued the improvement of the secondary defensive lines, conducted extensive training, and dispatched numerous security patrols throughout the regimental area.

[1] Unless otherwise noted, the material in this section is derived from: CG, 1stMarDiv, *Info for CG, FMFPac;* 1stMarDiv ComdD, Oct 52; 1stMarDiv PIRs 706–736, dtd 1–31 Oct 52; 7thMar ComdD, Oct 52; LtCol Robert D. Heinl, Jr. memo to Dir, MarCor Hist, HQMC, dtd 28 Oct 52, Subj: Notes on 7th Marines' Action (Defense of "The Hook"), 26–27 Oct 52, hereafter Heinl, *memo.*

These routine reserve roles were in addition to the primary mission of augmenting units on the Marine MLR in order to counterattack and defeat any attempted penetration of JAMESTOWN in the division area. As part of its counterattack mission, the divisional reserve regiment was to be prepared for employment anywhere in the I Corps sector to block an enemy advance.

On the division right, the 7th Marines remained on position in defense of JAMESTOWN. Following the bitter outpost contests on 6 October, Colonel Moore continued to retain all three battalions on line: 2/7 on the left, 3/7 in the center, and 1/7 on the right. The regimental commander had found it necessary to commit his three battalions on line due to the vastly overextended six-mile front, the rugged terrain, and the very real possibility of a major Communist attack anywhere along the MLR. With all battalions forward, Colonel Moore was left with a very small reserve, one company from 3/7. This battalion had to use as its reserve what had become known as "clutch platoons"—units composed of cooks, bakers, clerks, motor transport, and other Marine headquarters personnel. These local reserves, and even the reserve company from 3/7, could be employed only with the regimental commander's approval.

Line JAMESTOWN, in the 7th Marines area, meandered from the vicinity of the burned-out village of Toryom, on the left, to the Hook salient in the right battalion sector and from there southeast to the Samichon River, the boundary with the 1st Commonwealth Division. From the left battalion sector to the right, the terrain gradually grew more rugged until the hills finally spilled over into the Samichon Valley. To the rear of the MLR, the ground was less jagged; forward of the line, the hills were more precipitous in character. The steepest heights were in the right battalion sector. Highest terrain feature along Colonel Moore's MLR was Hill 146, located not far from the Hook. Throughout the 7th Marines sector rice paddies covered the narrow valley floors between the hills. Vegetation was sparse. A series of dirt roads and trails served the regimental area.

Combat outposts varied greatly as to their distance from JAMESTOWN. Farthest from the line were the three in the left battalion sector, manned by Lieutenant Colonel Caputo's 2/7. This trio, Carson, Reno, and Vegas, were approximately 1,000 yards forward of the MLR. Berlin and East Berlin (a new outpost established on

13 October) were the forward positions in the center line outposted by Lieutenant Colonel Charles D. Barrett's[2] Marines. To the right Lieutenant Colonel Dulacki's 1/7 sector held three—COPs Ronson, Warsaw, and Verdun, the latter near the Commonwealth border.

Ronson was the outpost nearest to the Hook, a major defensive position of the regiment. The importance of this part of the MLR, in the extreme eastern sector, lay not in its strength but rather in its weakness. Jutting as it did towards the Communist lines, the salient formed a J-shaped bulge in the main line, which not only gave the Hook its nickname but also established the vulnerability of the position. Its susceptibility to capture derived both from violation of a defensive axiom that the "MLR should not have sharp angles and salients"[3] and to the fact that the ridgeline on which the Hook was located continued northwest into Communist-held territory. Seattle, which the Chinese had seized on 2 October, lay only about 500 yards northwest of the Hook.

In spite of its vulnerability, the Hook could not be abandoned. There was no other terrain feature held by the Marines that could command the critical Samichon Valley, a major avenue of approach from the northeast directly to Seoul. The salient also dominated the entire nearby area of the Imjin River to the south. Possession of the Hook and adjoining ridge would give the Communists observation of a substantial portion of the Marine rear areas beyond the Imjin, as well as the vital river crossings. In the opinion of Major General M. M. Austin-Roberts-West, whose 1st Commonwealth Division was soon to take over the Hook sector, had the salient been lost, "a withdrawal of 4,000 yards would have been necessary."[4]

At the beginning of October, this vital area had been protected by COPs Seattle and Warsaw. When the former was overrun, it became necessary to establish a new position. This was directed by Lieutenant Colonel Dulacki, and on 16 October Ronson was estab-

---

[2]Responsibility for this part of the 7th Marines line changed on 13 October, when Lieutenant Colonel Barrett took command of 3/7 from Lieutenant Colonel Russell. The latter then was assigned as division senior liaison officer to the KMC regiment.

[3] Heinl, *memo.* The originator of this memo, Lieutenant Colonel Robert D. Heinl, Jr., was an experienced Marine officer and military historian who had just been assigned to the division for duty. Temporarily attached to the 7th Marines as an observer, his brief visit there happened to coincide with the beginning of the Hook battle.

[4] Quoted in LtCol Herbert F. Wood, *Strange Battleground: The Operations in Korea and Their Effects on the Defense Policy of Canada* (Ottawa: The Army Historical Section, Canadian Forces Headquarters, 1966), p. 213.

lished 200 yards southeast of Seattle and 275 yards west of the Hook. About 600 yards northeast of the salient the remaining position, COP Warsaw, commanded the lowlands to the east and the narrow, east-west oriented valley of a Samichon tributary immediately to the front.

Opposite the three MLR battalions of the 7th Marines were the 356th and 357th Regiments of the 119th Division, 40th CCF Army. In addition to these infantry units, numbering close to 7,000, an estimated 10 battalions (120 guns) of Chinese artillery[5] were facing Colonel Moore's regiment. Personnel strength of the American unit consisted of 3,844 Marines, 11 medical officers and 133 corpsmen, 3 U.S. Army communicators, and 764 Koreans (746 KSCs and 18 interpreters).

During the summer and early fall, the 7th Marines had amassed considerable information about the enemy, including Chinese strength and composition of forces and many of their combat characteristics. Encroachment on Marine ground positions by steadily creeping the CCF trenchline forward continued to be the enemy's major ground-gaining tactic. In fact, the Chinese units facing the Marine division concentrated their digging during the fall of 1952 in the sector north of the 7th Marines MLR. (See Map 15.) Other intelligence, however, seemed open to question. For example, there was the reported frontline presence of women among the 90 Chinese who had engaged a 2/1 patrol on 5 October as well as the sighting in the KMC sector on the 17th of enemy "super soldiers" far taller than the ordinary Chinese. Many in the division found it difficult to believe the statements of enemy prisoners. During interrogation they invariably maintained that the mission of Chinese Communist Forces in Korea was a "defensive" one.

The static battle situation encouraged the use of psychological warfare. In attempting to influence the minds of their opponents and weaken morale, the Chinese depended upon loudspeakers to

---

[5] The Marine division artillery regiment reported that in late October nine battalions of Chinese artillery, ranging from 75 or 76mm guns or howitzers to 122mm howitzers, opposed the 7th Marines. It was estimated that one other 122mm battalion was also emplaced north of the right division sector. In addition to these CCF units, elements of a 152mm self-propelled howitzer unit were also believed to be in the area. Late in November two batteries of 152mm howitzers were tentatively located about 4,000 yards west northwest of the Hook. Disposition had been determined "as a result of crater analysis, shell reports, sound plots, and capabilities of the weapon." 11thMar ComdD, Nov 52, "Enemy Artillery Activity Rpts," Nos. 21, 23, dtd 1, 21 Nov. 52.

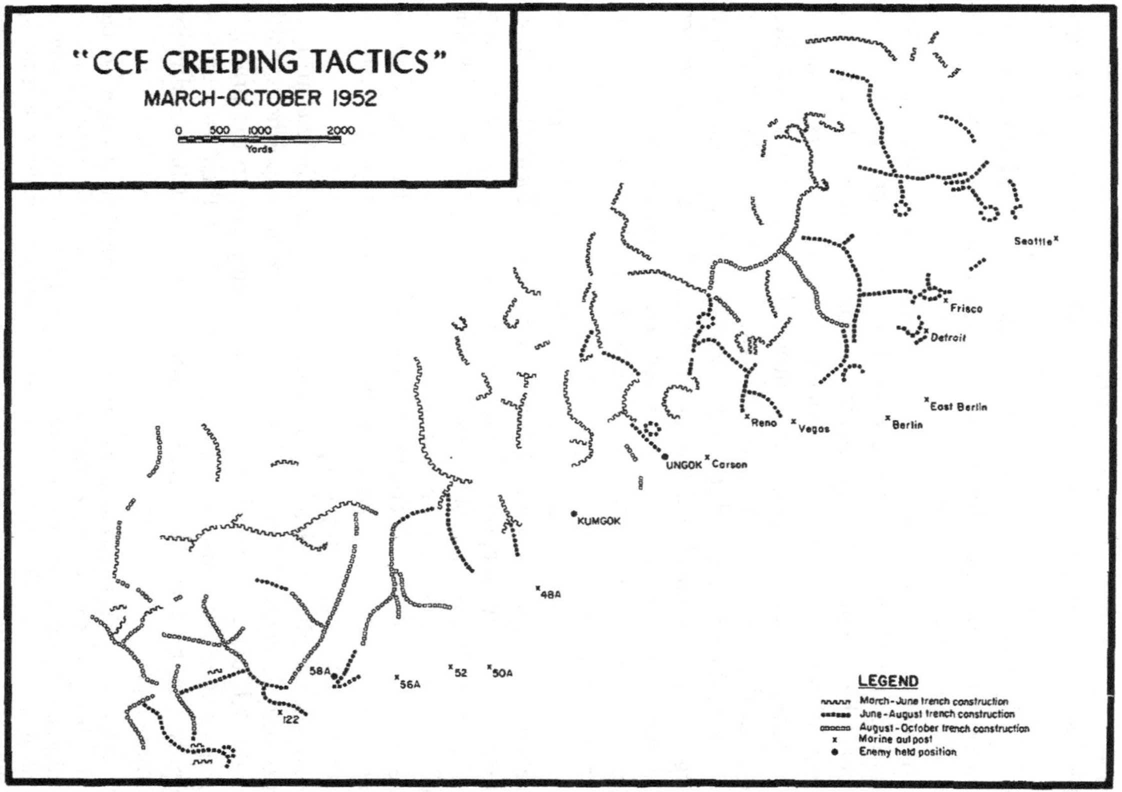
"CCF CREEPING TACTICS"
MARCH-OCTOBER 1952
0 500 1000 2000
Yards
Seattle
Frisco
Detroit
East Berlin
Berlin
Reno
Vegas
UNGOK
Carson
KUMGOK
48A
58A
56A
52
50A
122
LEGEND
March-June trench construction
June-August trench construction
August-October trench construction
Marine outpost
Enemy held position
K WHITE
MAP 15

carry their propaganda barrage across No-Man's-Land. Enemy employment of this technique was especially heavy during October. To Marines, for example, Chinese directed pleas of "Go home and have peace," "Surrender, we treat POWs well," "Leave Korea," "Marines, come and get your buddies' bodies," and the like, often to the accompaniment of music. On occasion, Chinese patrols left propaganda pamphlets behind them in the KMC sector. Infrequently, the enemy displayed signs along patrol routes urging Marines to surrender. Most of the Chinese psychological efforts were directed against the Korean Marines.

In enemy employment of artillery, Marine frontline units and division intelligence had become well aware of the vast improvements the Communists had made in recent months. Aided by a plentiful supply of ammunition, enemy guns and howitzers, including the heavy 152mm weapon, frequently delivered concentrated fires on critical positions in the division area. Marines felt the effects of how well the Chinese had learned to mass their fires against a single target for maximum destructive power. From the Marines, moreover, the enemy had picked up the artillery box tactic, employing it for the first time in their sector opposite Colonel Moore's regiment during the early October outpost battles.

During those same clashes, the 11th Marines had observed how the Chinese displaced some of their batteries well forward for more effective artillery support of their attacking infantry. One enemy artillery innovation had been noted the previous month by a Marine AO; on 19 September a Chinese artillery piece was detected firing in the open. Previous observations had indicated that the Chinese generally used wooded areas or extensive bunker-type positions to conceal their supporting weapons.

By the middle of October, 62.5 percent of the Chinese artillery opposing General Pollock's division was located in positions north of the 7th Marines. The importance the enemy put on the principle of massed artillery fire and the improvement of their ammunition supply can be seen in a remark attributed to a Chinese division commander:

> The enemy had organized an attack of two-battalion strength on our first-line platoon. As the enemy were getting into their assembly area I directed several volleys of rapid fire against them with a total expenditure of about 120 rounds. That very evening the army commander rang me up and said disapprovingly, 'You've expended a bit too much ammunition

today!' It seemed as though the army commander had detected precisely what was in my mind. There was an instant change in his voice as he said: 'Oh, comrade, it really could not be accounted as waste, but you must know we are short of supplies.'

Scarcely two years had passed but the situation was completely altered. In the present we had emplaced 120 guns to each kilometre of front line so that in a rapid-fire bombardment of 25 minutes more than 20,000 rounds of ammunition could be hurled against the enemy positions. If the fire used in supporting attacks and in repulsing enemy counter-attacks were taken into account the total would reach 70,000 rounds.[6]

Exaggerated as the numbers of guns and rounds may be, the basic massing technique was in line with U.S. intelligence estimates at the time. The remark also pointed to the importance the Chinese had learned to place on employment of artillery, a shift in emphasis that Colonel Moore's regiment was soon to experience in unprecedented volume.

Preparations for Attack and Defense[7]

Before the Hook battle erupted, the defensive fires that the 7th Marines could draw upon were not overpowering in terms of numbers of units available. Only one battalion, Lieutenant Colonel Bert Davis' 2/11, was in direct support of Colonel Moore's regiment. In this mission, the 2/11 fires were reinforced by those of 1/11 (Lieutenant Colonel David S. Randall). In addition to these organic units, the batteries of the 623d Field Artillery Battalion (155mm howitzers) and one platoon of C Battery, 17th Field Artillery Battalion (8-inch self-propelled howitzers) were readily available to the 7th Marines. In all, 38 light, medium, and heavy pieces constituted the artillery support of the right sector.[8] General support was available from Lieutenant Colonel Raymond D. Wright's 4/11 and from the 4.2-inch Rocket Battery (Captain Donald G. Frier). The 159th Field Artillery Battalion (155mm howitzers) and B

---

[6] *CPV, Recollections*, p. 360.

[7] Unless otherwise noted, the material in this section is derived from: 1stMarDiv ComdD, Oct 52; 1stMarDiv G-3 Jnls, 24-26 Oct 52; 1stMarDiv PIRs 729-732, dtd 24-27 Oct 52; 7th Mar, 1/7, VMA-323 ComdDs, Oct 52; Heinl, *memo.*

[8] 11thMar ComdD, Oct 52, App III, Sheet 3. Eighteen of the weapons (the 623d Field Artillery Battalion) had just moved into the Marine sector and begun operating on 14 October. The unit remained under I Corps operational control, with the mission of providing general support reinforcing fire.

Battery, 204th Field Artillery Battalion (155mm guns), like the other Army units positioned in the Marine Division sector, reinforced the fires of division artillery. Fire support from 1st Commonwealth Division weapons within range of the Hook area could also be depended upon.

Although the Army artillery units satisfied the heavy punch requirement of the 11th Marines, commanded since 21 September by Colonel Harry N. Shea, there was one basic element the regiment lacked. This missing ingredient was a sufficient amount of ammunition for the howitzers. Defense of outposts and mainline positions along the EUSAK front in early and mid-October 1952 consumed a great deal of this type of ammunition. This heavy expenditure was brought to the attention of the corps commanders by Eighth Army. General Van Fleet pointed out that ammunition consumption rates for both the 105mm and 155mm howitzers during these two critical weeks in October not only exceeded the expenditures of the massive Communist spring offensive in 1951 but also the UN counterstroke that followed.[9]

To help remedy the situation, the EUSAK commander urged "continuous command supervision to insure the maximum return for all ammunition expended."[10] The general made it plain that he was not changing his policy of exacting a heavy toll whenever the enemy began an attack. This course had been followed by the 1st Marine Division, but the Marines' ability to both restrict the enemy's creeping tactics and simultaneously fight a siege-type war was noticeably impeded.[11]

As the end of October approached, the shortage of ammunition was becoming a subject of increased concern to the frontline Marine units. Daily allowances established for the last 11 days of the month were 20 rounds of 105mm high explosive (HE) and 4.3 rounds of 155mm high explosive for each tube.[12] With such small quantities

---

[9] Later in 1951, during the UN Summer-Fall offensive, ammunition consumption had again risen sharply, creating concern among corps commanders and occasioning one of them to remark to a subordinate, "We have the distinct impression that two of your battalions are trying to compete for a world's record." Capt Edward C. Williamson, *et. al.*, "Bloody Ridge," ms OCMH, 1951, cited in James A. Huston, *The Sinews of War: Army Logistics, 1775–1953—The Army Historical Series* (Washington: OCMH, 1966), v. II, p. 632.

[10] 1stMarDiv ComdD, Oct 52, App I, No. 19.

[11] *PacFlt EvalRpt* No. 5, Chap. 8, p. 8–71.

[12] *Ibid.*

to fire and further restricted by an equally critical shortage of both hand grenades and 81mm mortar rounds, Colonel Moore was almost powerless to spike the Chinese preparations for assault of the Hook.[13] Artillery fires were reserved for only the most urgent situations or for large bodies of troops. It was one observer's opinion that the "enemy could show himself almost at will without receiving fire, and that it was impossible either to harass or neutralize his continual fortification activity, let alone embark upon systematic destructive fires of the kind he was carrying out."[14]

As a means of compensating for the shortage of 81mm mortar and 105mm howitzer ammunition, the Marines reverted to a former method of using machine guns. This technique, employed during the trench warfare days of World War I but seldom thereafter, was considered a useful expedient to discourage enemy defensive creeping tactics as well as to deter his preparations for objective attacks. The system required emplacing heavy machine guns both on and to the rear of the MLR to fire into areas that troops used for assembly or as check points. If the target was visible to the machine gunner, he could take it under direct fire. At night, when the enemy operated under cover of darkness, the machine guns fired into zones which had already been registered in the daytime. Colonel Moore directed his units on 23 October to resort to this expedient.

A 1st Marine Division daily intelligence report covering the 24-hour period beginning at 1800 on 24 October noted that there was "a marked increase in enemy artillery and mortar fire with an estimated twelve hundred rounds falling in the CT 1010 area of the 7th Marines sector."[15] According to the division PIR there was also an increased number of enemy troops observed that same day in locations west and northwest of the Hook. Most of the fire was directed against the Hook area of the MLR and on the two sentinels, Ronson and Warsaw. Efforts by Marines and some 250 KSCs to repair the damaged or destroyed bunkers, trenches, communications lines, and tactical wire, during brief periods of relief from the artillery deluges, were wiped out again by subsequent shellings.

---

[13] For example, during the latter part of the month each rifle company in the Hook battalion was limited to 150 hand grenades. The total 11-day allowance for Lieutenant Colonel Dulacki's 81mm mortars was 475 rounds. 1/7 ComdD, Oct 52, App. III.

[14] Heinl, *memo.*

[15] 1stMarDiv PIR 729, dtd 24 Oct 52, p. 2. Ronson, the Hook, and Warsaw are within the 1,000-meter square, CT 1010.

It would not be correct to say that 1/7 remained entirely passive at this time. Battalion weapons replied, though in faint voices barely audible in the din created by Chinese firing. Regimental mortars chimed in and so did 2/11, which fired 416 rounds in the 24 hours ending at 1800 on the 24th. For that same period, tanks expended 137 rounds at active weapon positions firing on the Hook. One air strike was directed against the enemy opposing the Hook battalion. This attack by a quartet of Marine F9Fs from VMF–311 (Lieutenant Colonel Arthur H. Adams) bombed and napalmed a troublesome group of Chinese entrenched on the enemy MLR 750 yards east of the Hook.

During the next 48 hours, the enemy continued his preparations for an attack, concentrating his artillery fire on the Hook area. Colonel Moore's battalions received approximately 2,850 artillery and mortar rounds, most of which rained down on 1/7 to the right. There, the heavy and continuous fire slowed Marine efforts to restore their wrecked bunkers and trenches. Late on the 25th there was some relief from the artillery bombardment, but by that time many of the prophets on the line and in the rear area were uncertain only as to the precise time of the unexpected Chinese attack.

Colonel Clarence A. Barninger, the division intelligence officer, had himself alerted General Pollock to the implications of "the intensification and character of enemy fires"[16] being received in the 1/7 sector. The intelligence evaluation was not based only on recent events. A detailed study of Chinese capabilities and possible courses of action had just been completed by the G–2 and his staff. In its discussion of the early October outpost attacks in the division right, the report concluded that Chinese interests lay in gaining the "terrain dominating the Samichon Valley. . . ."[17]

Since 5 October when 1/7 had been moved into the line as the regiment's third MLR battalion, the enemy had begun a regular shelling of 1/7 positions adjacent to the Hook. Incoming rounds had increased almost daily. "Troops, vehicles, and tanks moving in daylight even behind the MLR almost invariably brought down enemy artillery or mortars upon them. It was apparent that the enemy was making preparation for a large scale assault in this

---

[16] Heinl, *memo.*

[17] 1stMarDiv Intell. Est., dtd 19 Oct 52, p. 8, filed with the divisions PIRs for that month.

portion of the MLR,"[18] the battalion commander, Lieutenant Colonel Dulacki, later recalled. Matters took an even more ominous turn about 23 October when the Chinese "began a deliberate, deadly accurate precision fire aimed at destruction of the major fortifications in the Hook's system of dug-in defense."[19] As the tempo of this fire stepped up daily, the destruction of the battalion's carefully prepared defenses exceeded the Marines' ability to repair the damage. The artillery build-up was believed preparatory to an attempt to either seize or breach the MLR.

In late October, Lieutenant Colonel Dulacki had two companies on the MLR to protect this important area. On the 23d, Captain Frederick C. McLaughlin's Company A was assigned the left part of the battalion sector, which included the Hook. A squad outposted Ronson and a reinforced platoon was stationed at Warsaw. At 0200 on the 26th, Company C (Captain Paul B. Byrum) departed the battalion reserve area to take over responsibility as the left MLR company. Relief of Company A was completed at 0410.[20] Holding down the right flank of the main line during this time was Company B (Captain Dexter E. Evans). This area was larger but somewhat less rugged than the western part of the 1/7 sector.

In the two days immediately preceding the Chinese attack of 26 October, 1/7 received a limited amount of support intended to harass the enemy and throw him off balance, if possible. Tanks fired their 90s at bunkers, caves, trenches, and direct fire weapons in the enemy sector. On the 25th, Company A of the 1st Tank Battalion blasted away 54 times at these targets; on the next day, Captain Clyde W. Hunter's gunners more than tripled their previous day's output, firing 173 high explosive shells. Artillery, in the meantime, stepped up its rate of fire on the 25th, when Lieutenant Colonel Davis' 2/11 fired 575 rounds, followed by 506 more the next day. The division general support battalion, 4/11, fired a total of 195 rounds on these two days.[21] Nearly half were to assist the 7th

---

[18] *Dulacki ltr.*

[19] Heinl, *memo.*

[20] Due to the width of the Hook sector, it was necessary to keep all three rifle platoons in the line. A reinforced platoon from the battalion reserve outposted Warsaw. While Company A was on line, a Company C platoon manned the outpost; when Company C was relieved on 26 October, a Company A platoon was sent to Warsaw. Maj Frederick C. McLaughlin ltr to Dir, MCHist, HQMC, dtd 27 Jan 70, hereafter *McLaughlin ltr.*

[21] On 24 October, Battery M of the battalion was temporarily relaid to provide additional support to Colonel Moore's regiment.

Marines. On both days the regiment received the benefit of 4.5-inch rocket ripples.

Air support just prior to the attack was increased slightly, but only two strikes were flown for the Hook battalion. At 1535 on the 25th, two Corsair fighters and a pair of AUs, the attack version of the Corsair, dive-bombed a section of Chinese trench that housed a number of weapons bothersome to the Marines nearby.[22] The four VMA-323 aircraft claimed destruction of 40 yards of trench and damage to 35 yards more. The target was 1,000 yards southwest of the Hook. Next morning the squadron sent three of its famed fighters against bunker positions on a hill 900 yards west of the 1/7 salient. This mission had been prebriefed to attack enemy artillery positions opposite the KMC line. Instead, the flight was diverted to take on the bunkers, which represented, at that time, more of a menace to the division. The attack destroyed one bunker, damaged another, and produced an estimated seven casualties.

Hidden nearby the area of this air strike in the early morning hours of 26 October was the Chinese infantry unit which later that same day would attack the Hook. Before daybreak the 3d Battalion, 357th Regiment, had moved from an area nearly two miles west of the Hook. The forward elements, two companies, with two day's rations for each man, halted about a mile from their objective. There the Chinese remained throughout most of the 26th, carefully concealing themselves from observation by friendly forces.[23] While the enemy troops were lying low, their mortars and artillery began the final preparatory fires.

---

[22] The flight had been scheduled to attack active artillery positions 3½ miles north of the Carson-Reno-Vegas area. When some of their ordnance was unexpended after putting these guns out of action, the planes, were ordered to take on the trench target.

[23] Within the division there were no reports of sightings of unusually large groups of enemy soldiers in this area. In fact, there were fewer enemy seen on the 26th than any other day since 18 October. During the 23d and 24th, about 100 enemy had been observed almost a half mile closer to the Hook than the hideout area used on the 26th. 11thMar ComdD, Oct. 52, p. 12; 1stMarDiv PIR 729, dtd 24 Oct 52, p. 2.

Attack on the Hook[24]

On the morning of 26 October, Chinese artillery and mortar fire striking the MLR slackened a bit but was still sufficiently heavy in the vicinity of the Hook to prevent visitors in the area any direct observation from the salient. During his inspection of Hook defenses that morning, Lieutenant Colonel Dulacki was knocked to the ground by the concussion of an enemy artillery round exploding nearby.[25] In the afternoon, enemy shelling continued at a steady pace, but towards the end of the day intense mixed artillery and mortar fire increased to preattack proportions. Dusk brought no relief from the enemy's supporting weapons.

Out at the flanking positions, Ronson and Warsaw, there was little change in the intensity of the enemy shelling for the remainder of the afternoon. Bunkers and trenches were caved in, just as they were on the Hook[26] from the preparatory fires that had been building up over a period of days. (For a sketch of the Hook battle area on 26 October, see Map 16.) Enemy shelling had also produced a number of casualties. Marines at Ronson were the first to experience the enemy's ground assault. At 1810 the outpost reported an increased rate of mortar and artillery rounds exploding on the position. Two groups of enemy soldiers were seen moving towards the outpost, one from the east and the other from the west. Ronson Marines took these advancing soldiers under fire immediately.

Initially, the radio messages from Ronson reported that the attacking force was a company, but a later estimate of approximately 50 Chinese appeared to be more nearly correct. Communist infantry

---

[24] Unless otherwise noted, the material in this section is derived from: 1stMarDiv ComdD, Oct 52; 1stMarDiv G–3 Jnl, 26 Oct 52; 1stMarDiv PIRs 723, 734, dtd 27, 29 Oct 52; 7thMar, 11th Mar, 1/7, 4/11, 1st TkBn, VMF(N)–513 ComdDs, Oct 52; Heinl, *memo*.

[25] The 1/7 commander, who was uninjured by the blast, might have become a believer that day in the military cliche, "Rank hath its privileges," for Brigadier A. H. G. Ricketts (29th British Infantry Brigade, 1st Commonwealth Division), who was standing near Lieutenant Colonel Dulacki, was untouched. The British division was scheduled to take over responsibility for the Hook sector in early November.

[26] Prior to the enemy's steady shelling of the Hook, the trenches were six feet deep. The preparatory fires of the past several days had been so intense that in nearly all areas the trenchline had been leveled by the time of the Chinese attack. "I am convinced that the Chinese didn't realize that they had penetrated our MLR or they would have exploited the penetration." Col Russell E. Honsowetz ltr to Dir, MCHist, HQMC, dtd 26 Jan 70.

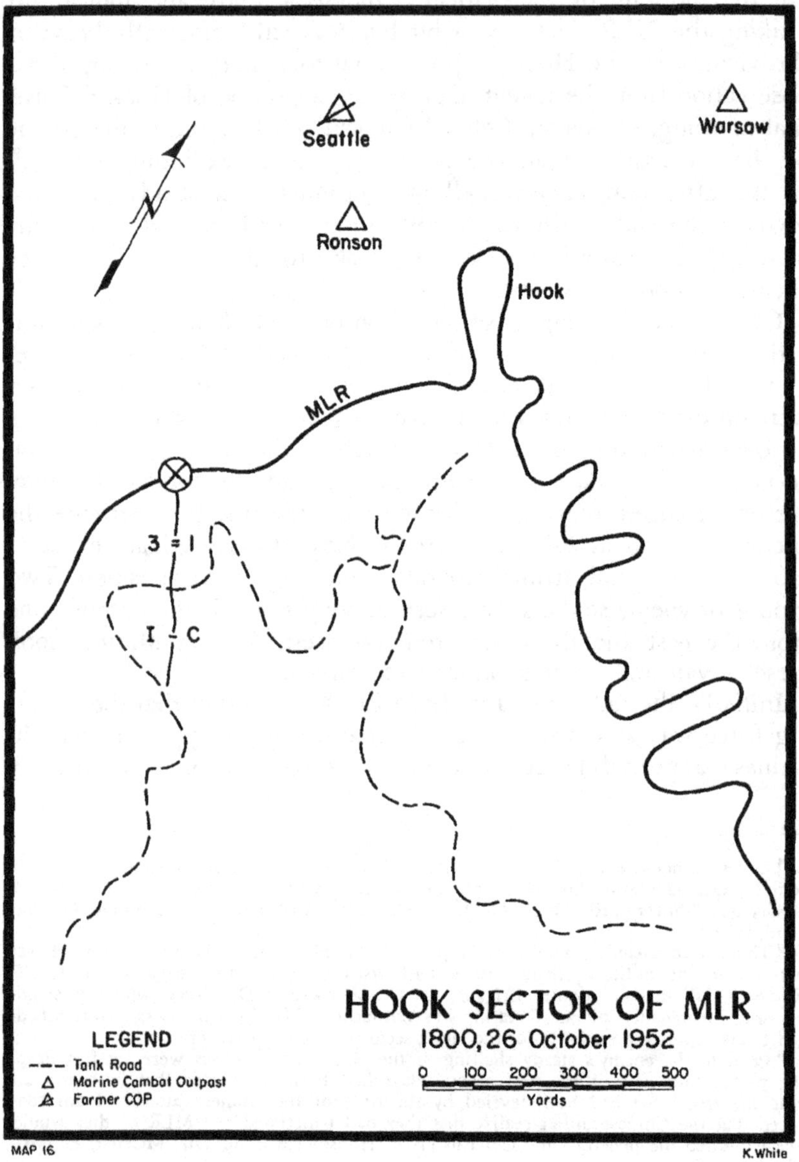

MAP 16

K. White

made their way through the defensive artillery barrages requested by the COP garrison and into the rifle and machine gun fire of the Marines. By 1838 the enemy had overrun the squad of Marines and was in possession of Ronson. No one had escaped from the outpost.

At this time, 800 yards northeast, the 9th Company, 357th Battalion was working its way towards Warsaw. Striking at the COP from both east and west, the enemy company was momentarily halted by extremely heavy Marine mortar and artillery fire. By 1820, the platoon at Warsaw had requested the protective box around its position; this fire the 11th Marines delivered promptly. Still the Chinese continued to besiege the position and Company A defending Marines, under outpost commander Second Lieutenant John Babson, Jr., were locked in a hand-to-hand struggle. As a platoon was being readied to reinforce Warsaw the outpost reported, at 1907, that enemy soldiers had reached the Marine bunkers and that the defenders were using bayonets, pistols, hand grenades, and both ends of their rifles to repel the Communist invaders.

Three minutes later came the word, "We're being overrun." With this message all communication from the outpost temporarily ceased, but at 1944, Lieutenant Colonel Dulacki's CP heard Warsaw report heavy fighting still in progress there. The outpost first stated that enemy soldiers were on top of the bunkers; then called for "VT on own position" which the 11th Marines furnished.

The seriousness of the situation was immediately apparent at higher commands. One outpost had been lost; a second was in jeopardy. At about this time, a veritable avalanche of enemy artillery and mortar fire began to blanket the Hook. Colonel Moore released Captain McLaughlin's company to 1/7. The 7th Marines commander also ordered regimental ammunition supplies be allotted to Lieutenant Colonel Dulacki's area. Shortly after that, division lifted ammunition restrictions on 1/7.

To counter the impending ground attack, at 1859 Lieutenant Colonel Dulacki ordered Captain McLaughlin's Company A forward to reinforce the Hook sector and to assist Company C in containing the enemy attack. One platoon, the 1st, departed immediately for the MLR. As the remainder of the company prepared to move out, the enemy struck in estimated battalion strength. By 1938 some of the CCF infantry had advanced to the main trenches immediately south of the Hook. Within a few minutes, a second wave of Communist soldiers, following closely the preparatory barrages, hit JAMESTOWN just east of the 1/7 salient and frontally at the Hook itself. It appeared that the Communists had come to stay, for many cargo carriers—Chinese with construction materials for bunkers and trenches—accompanied the attacking infantry.

Fire fights raged during the early phase of the struggle, with continuous support furnished the assault troops by Chinese artillery and mortars. The momentum of the enemy's three-pronged attack, aided by heavy rear area fire support, enabled the Chinese to overrun the trenches and push on along the crest of the ridge, its slope near the spine, and across the segments formed by the spurs that jutted south from the crest. Marine defenders pulled back while a small rear guard covered their movement with fire. Along the MLR, about 400 yards south of the Hook, the Chinese had slipped around the flanks of the COP and at 2030 forced a penetration in the C/1/7 line. Second Lieutenant John W. Meikle (1st Platoon, Company C) organized the Marines into a perimeter defense adjacent to the MLR. At 2130, remaining elements of the company formed another defense blocking area 550 yards east of the Hook near the crest of the ridge.

Between these two positions small groups of Marines continued the heavy close fight to repulse the enemy while inching their way forward to tie-in with the rest of the unit. (See Map 17 for penetration limits during the Hook battle.) To the northeast, the platoon at Warsaw had not been heard from since 1945, and at 2330, Colonel Moore reluctantly declared the outpost to be in enemy hands.

At the time the loss of Warsaw was announced, counter-measures designed to halt the enemy assault were in various stages of preparation or completion. The initial reinforcing element sent forward to strengthen the main line had linked up with Lieutenant Meikle's 1st Platoon, Company C, in the perimeter near the 3d Battalion boundary. The remainder of Company A was en route to the crest of the east-west ridge to thwart what appeared to be the main enemy drive. Colonel Moore had released his meager reserve, H/3/7, at 0300 on the 27th, and General Pollock had ordered one of the division reserve battalions, 3/1, to the 7th Marines area, although still retaining operational control of the unit.

As the forward battalion of the division reserve, 3/1 (Lieutenant Colonel Altman) had prepared counterattack plans for critical loca-

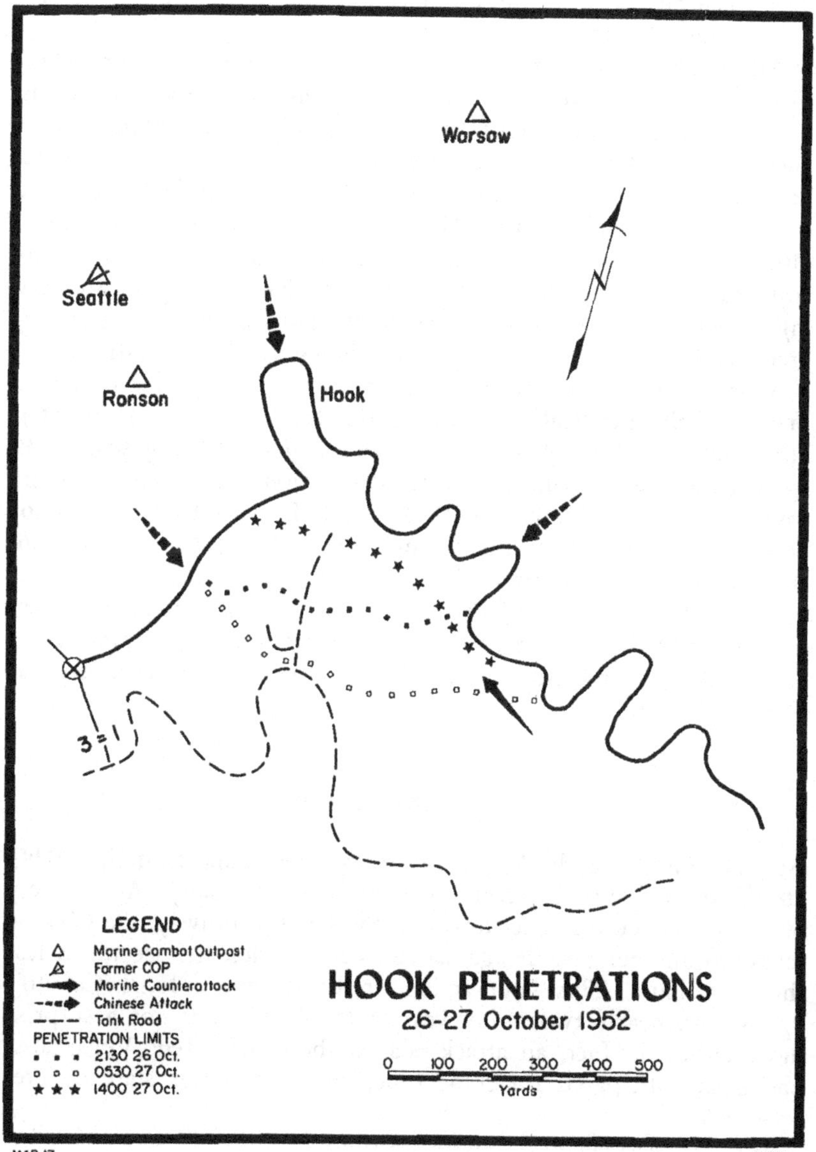

MAP 17

K. White

tions in the division sector and had previously made a reconnaissance of the Hook area. The battalion immediately displaced from its bivouac site north of the Imjin (Camp Rose) to an assembly area behind the 7th Marines on the MLR.

All possible support for 1/7 was made available, since the critical situation resulting from the major enemy assault automatically suspended previous restrictions on use of artillery and mortar allowances. At Warsaw, 2/11 blanketed the position with a continuous barrage in order to limit the enemy's ability to effectively hold and consolidate the captured COP. Lieutenant Colonel Davis' cannoneers also blasted enemy formations in response to fire missions from forward observers. Artillery rounds fell on Chinese outposts supporting the attack, on approach routes to the battleground, on assembly areas, and on known and suspected Chinese artillery locations.

Marine aviation and tanks were employed as part of the plan to first limit the penetration made by the enemy before the counterattack to expel him. A section of tanks had been firing since 1930 against the enemy main line; a second section joined the direct fire assault a half hour later. At 2113, one F7F, with 1,300 pounds of bombs, hit a portion of the enemy's MSR. At 2306, another twin-engine Grumman Tigercat blasted the same area, about three-quarters of a mile west of the Hook. These initial one-plane strikes in support of the defense of the salient were flown by Captain Leon C. Cheek, Jr. and Major Laurel M. Mickelson, respectively, of VMF(N)-513.

Reno Demonstration[27]

At 0030 on the 27th, Major Mickelson, returning from his MPQ attack, touched his Tigercat down at K-8 (Kunsan). At the very moment that the plane set down on the Kunsan runway, the Chinese launched another assault against the 7th Marines, the second in less than six hours. This later action, in Lieutenant Colonel Caputo's 2/7 sector, nearly two miles west of the Hook, was not a surprise move either. In fact, an attack against the Carson-Reno-Vegas area had been anticipated for some time, and it was this state of pre-

---

[27] The material in this section has been derived from 7thMar, "Summary of Action, 26 Oct–1 Nov 52, Hook, Reno, Ronson"; 2/7 ComdD, Oct 52.

paredness that throttled the enemy's attempt to seize an outpost here.

Division intelligence had accumulated considerable evidence that the Chinese buildup in late October was intended to ultimately clear the way to the 2/7 outposts rather than those of 1/7 in the eastern Hook area. A majority of the Marine supporting arms effort immediately prior to 1800 on the 26th had gone to the left battalion of Colonel Moore's regiment. Aware of the interest the enemy had shown in the outposts earlier in the month, the battalion commander had strengthened the defense of this key area. One measure, increasing the size of the ambush force maintained at night near Reno from a squad to a platoon, was to pay handsome dividends before October was over.

Just after dark on the 26th, a reinforced platoon from Captain James R. Flores' Company E departed the MLR on a combat patrol and ambush mission. After reaching its assigned area, about 300 yards short of the hill that housed COP Reno, the ambush platoon disappeared into camouflaged dug-in positions and waited. At midnight, the Marines were alerted by faint noises to the front. There, elements of two Chinese companies, which had stealthily maneuvered into the ambush area, were organizing for a sneak assault by an envelopment on Reno from the rear. (See Map 18.) The waiting platoon apprised the outpost of the enemy's presence in the area; then when it appeared that the Chinese were about to launch their assault, the ambushers opened fire.

As the surprised Chinese turned to take on the hidden ambush platoon, the two defending squads at Reno began firing. It took 10 minutes before the Chinese were sufficiently recovered to organize a withdrawal. At 0040, enemy elements quickly began to pull back towards the north. The outpost had been spared a major action, but its occupants were to be again engaged by the Chinese before daybreak.

At 0400, one platoon from a third CCF company, approaching from an enemy hill to the northeast, hit Reno. The attack was conducted in a fashion not previously experienced by the 1st Marine Division in West Korea—platoons echeloned in depth, assaulting in successive waves. The first unit to reach Reno was composed of grenade throwers and supporting riflemen. This advance element was followed immediately by the rest of the platoon, infantry armed with submachine guns and rifles. Marines on Reno were not troubled

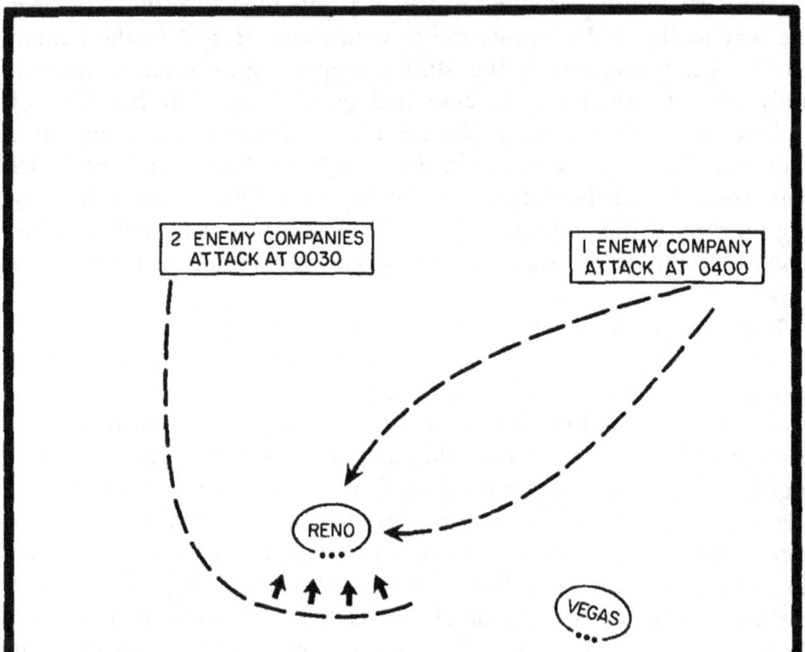

MAP 18

K.White

by the initial platoon asault, but the second one made some inroads before the defenders' fires forced the enemy to pull back. A third two-phased attack succeeded, however, in cracking the defenses at the northeast section of the position. The outpost commander then ordered his Marines into the bunkers and called for overhead artillery fire. Caught in the open, the Chinese were forced to withdraw at 0440 and did not return.

Counterattack[28]

After the Marines in Lieutenant Colonel Caputo's 2/7 sector had dealt with the demonstration force, the action shifted back to the Hook. Early on the morning of the 27th, Captain McLaughlin's unit, sent to the Hook-Hill 146 crest to block the penetration of the MLR, had established contact with Captain Byrum's Company C, passed through its lines, and pressed on to the Hook. Suddenly, enemy small arms and machine guns opened up on lead elements of Company A. Artillery and mortar fire then began to hit the company. The Marines continued their advance and made some progress in arresting the Chinese thrust at the ridge. Shortly thereafter the enemy called in heavy supporting fires, forcing Company A to halt its attack temporarily. When the company commander ordered his men to resume the advance, overwhelming enemy fire again slowed the movement. McLaughlin then ordered his men to hold and dig in.

When report of the Company A situation reached the regimental CP, Colonel Moore ordered into action his last reserve unit, Captain Bernard B. Belant's Company H.[29] He was directed to report to 1/7, then to pass through the depleted ranks of Company A, and take up the attack downridge towards the salient. At 0340 the regiment attached H/3/7 to 1/7 for operational control; at 0505 the company arrived at the 1st Battalion CP. Forty minutes later, Company H reached Captain McLaughlin's area, where it regrouped and then deployed toward the ridgeline for the counterattack.

---

[28] Unless otherwise noted, the material in this section is derived from: 1stMarDiv ComdD, Oct 52; 1stMarDiv G–3 Jnl, dtd 27–28 Oct 52; 7thMar, 11thMar, 1/7, 4/11, 1st TkBn, VMAs–121, –212, –323 ComdDs, Oct. 52.

[29] At 0545 on the 25th, Company I (Captain John Thomas), then the regimental reserve, and Captain Belant's Company H, responsible for the right sector of 3/7, had exchanged roles.

When Captain Belant led his Marines towards the Hook to oust the Chinese, the enemy drive had reached the point of its deepest penetration. By this time the Chinese had seized control of slightly more than a mile of the meandering MLR. Most of the captured main defense line extended from the Hook east along the ridge towards Hill 146. (One-third of the Communist advance was from the Hook southwest, in the direction of the 3d Battalion boundary.) Between 0545 and 0800, H/3/7 worked its way towards the Hook-Hill 146 crest. After two hours the company was at the ridgeline, and at 0800 Captain Belant was ready to move forward towards the salient, a straight-line distance of about a half-mile. On the hour, the push downridge started. After having advanced about 200 yards, the H/3/7 Marines were assailed by small arms fire and the rain of heavy caliber rounds supporting the enemy's thrust. Captain Belant signalled his Marines to attack.

Immediately, Second Lieutenant George H. O'Brien, Jr. leaped up from his position and shouted for his platoon to follow. On the run, he zigzagged across the exposed ridge and continued down the front slope towards the main trench. Before reaching this objective, the platoon commander was knocked to the ground by the impact of a single bullet. Scrambling quickly to his feet he motioned for his men to follow and took off on the run for the enemy-occupied trenchline. Again he stopped, this time to assist an injured Marine.

As he neared the trenchline, Lieutenant O'Brien started to throw a hand grenade into the enemy-occupied bunkers, but was stopped by the Chinese. With his carbine, the officer methodically eliminated this resistance, then hurled the grenades. Overcoming this position, the Texas Marine and his platoon advanced towards the Hook, but the enemy, now partly recovered, was able to slow and ultimately stop the counterattack. A profusion of artillery and mortar fire was primarily responsible for halting the advance, which had carried Company H very close to the Hook bunkers.

Spurred on by the leadership of Lieutenant O'Brien, who later received the Medal of Honor,[30] the company was able to execute

---

[30] Another Medal of Honor resulting from the Hook action was awarded posthumously to Second Lieutenant Sherrod E. Skinner, Jr. for "conspicuous gallantry and intrepidity." Lieutenant Skinner, whose twin brother was also a Marine officer, had been assigned as an artillery forward observer with F/2/11. When the Chinese attack hit the MLR, Lieutenant Skinner organized the surviving Marines in defense of their observation post. Fighting off the enemy and calling down defensive artillery fire on the

a limited advance. Despite the heavy artillery and mortar fire, the company drove a wedge into the Communist position, thereby retaking the initiative from the enemy. Company H also took three prisoners in the southeast end of the Hook before being forced by a deadly enemy mortar and artillery barrage to withdraw upridge.

The attack by Company H had been well supported from the air. At 0840, a flight of four ADs from Lieutenant Colonel Cargill's VMA–121 assaulted the former Marine COP Seattle, where enemy reinforcements were being funneled through on the way to the Hook. Bombs and napalm took a heavy toll of the troops, bunkers, and weapons pouring fire on the counterattacking Marines. One hour later, a division (four planes) from VMA–323 struck another trouble spot, a former Marine outpost known as Irene (later, Rome). Aircraft of Lieutenant Colonel Chamberlain's squadron hit this objective with three tons of bombs and more than 4,000 pounds of burning napalm. Thirty minutes later, another foursome, these from VMA–212, (Lieutenant Colonel Charles E. Dobson, Jr.),[31] delivered bombs, napalm, and 20mm shells on enemy soldiers moving on the MSR towards JAMESTOWN.

While these three squadrons were bombing enemy strongpoints and other targets of opportunity, division artillery and tanks continued their destructive fire missions. Between 0930 and 1300, two tanks from Company A, 1st Tank Battalion, blasted away at Chinese bunkers and trenches, at an enemy 76mm gun on Seattle, and at positions southwest of the Hook. Artillery—2/11, 4/11, and the rocket battery—contributed the weight of its support. The 11th Marines, in an effort to stop the heavy hostile shelling of the Hook sector, fired 60 counterbattery missions on Chinese gun emplacements during the first 24 hours of the attack.

In the early afternoon of the 27th, 1st MAW attack squadrons continued their bombing and strafing of enemy troops engaged in

---

assaulting Chinese, he delayed capture of the position. Twice he left the bunker to direct fire on the enemy and get more ammunition.

When the Communists finally overran the bunker, Lieutenant Skinner instructed his fellow Marines to pretend they were dead; during the next three hours several different enemy groups frisked the inert Marines without discovering their ruse. Later, when a skeptical enemy soldier hurled a grenade into the bunker, Lieutenant Skinner unhesitatingly rolled on top of the missile, shielding the two surviving Marines. By thus absorbing the full force of the explosion, he sacrificed his life for theirs. (2dLt Sherrod E. Skinner, Jr. Biog. File)

[31] The new squadron commander had relieved Lieutenant Colonel Maurice Fletcher two days earlier. This flight was the first of two CAS attacks in behalf of the Hook forces that the new commanding officer participated in that day.

the assault against the Hook. Before sundown, 30 aircraft had taken part in 8 additional strikes in support of Marine counterattacks along the ridge. The number of aircraft involved in close air support sorties for the Hook was approximately half the number received by the division all day. Of the 72 aircraft flying CAS strikes during the first 24 hours of the Hook action, 67 were Marine planes, all from MAG–12.

As in the morning's close air support flights, Lieutenant Colonel Cargill's ADs provided the bulk of air support for ground action that afternoon. Striking first a command post southeast of the 1/7 salient, at 1410, VMA–121 came back a half-hour later with four more Skyraiders against CCF troops pressing to envelop the right flank of the counterattack force. At 1635, two squadron aircraft flew in quickly in response to a sighting of troops moving forward in the Samichon tributary 1,000 yards north of the Hook. Twenty minutes after this successful attack, four more Skyraiders attacked bunkers opposite the left flank of Lieutenant Colonel Dulacki's sector. The final daylight strike for 1/7 was again made by four ADs from –121. These planes took under attack a target that had been bombarded just 25 minutes earlier by Corsairs from VMA–323.

Another Marine attack squadron, VMA–212, participated in the Hook support that afternoon. At 1344, a four-plane flight assaulted troops moving through Frisco to reinforce the Chinese drive on the Hook. Two of the planes dropped three 1,000-pound bombs and two 250-pounders on the enemy soldiers. The other pair of attack Corsairs released six 780-pound napalm tanks over the position. It was estimated that 25 Chinese casualties resulted from this air attack. Wrapping up the VMA–212 CAS for the Hook sector on the 27th was a strike, at 1440, on camouflaged positions and another at 1520 against caves and bunkers. Each of these air assaults took place about 950 yards from the Hook. The earlier one was a napalm attack from 50 feet above the ground. One of the six tanks would not release and three did not ignite. Four caves were destroyed and one bunker was damaged in the latter strikes.

Between the morning and afternoon air strikes, the ground commanders put together the final plans for recapture and defense of the Hook. When General Pollock had released I/3/1 to the regiment during an inspection trip to the 1/7 area that morning, the company was already en route to the ridge to make the counterattack. The ground commanders agreed that after I/3/1 regained the salient, H/3/1 would take over the right sector of 1/7 and the relieved company, B/1/7, would then occupy both the critical MLR sector and Warsaw. Lieutenant Colonel Dulacki's scheme to recapture the positions and ground lost on 26 October was a continuation of the attack from atop the ridge directly towards the objective. It was to be a hard-nosed, frontal assault, but the only maneuver deemed advisable.

Clearing the Company C command post about noon, the lead elements of Captain Murray V. Harlan, Jr.'s Company I, the 1st Platoon, continued its route to the ridge. After the 40 Marines had gained the crest, they quickly reoriented themselves to the new direction, and at 1350, led the I/3/1 assault. Artillery preparation by the 11th Marines had preceded the crossing of the line of departure, and these supporting fires were partially responsible for the substantial initial advance made by the counterattacking Marines. But Chinese artillery was not idle at this time either, and the volume of enemy fire matched that of the Marines. The I/3/1 movement forward was also slowed by Communist soldiers, estimated at about a company, who fired from protected positions along the perimeter of the Hook.

Inch by inch the company crawled forward. The vicious Chinese supporting barrages were exacting many casualties among Captain Harlan's troops,[32] yet they crept on, and ultimately reached the artillery forward observer bunker atop the ridge but 150 yards short of the Hook trenches. At this time, 1635, the enemy supporting fires were directed not only on the advancing Marines and the MLR defenses but extended as far back as the regimental CP.[33] Chinese soldiers still clung to some of the Hook positions and trenches of the MLR just below the crest on the northern sides. Marines closest to the Hook could see the virtual ruination caused by enemy artillery and mortar shells to the trench system within the salient.

Nearing their objective, elements of Company I pressed on with even more determination. By 1700 a few had made it to the shell-torn ditches, where they sought momentary refuge to reorganize. Several more joined, and together they reconnoitered the trenches

---

[32] During this action, the company suffered 15 killed, 71 seriously wounded, and 6 slightly wounded. 3/1 ComdD, Oct. 52, p. 3.

[33] 1stMarDiv G-3 Jnl, dtd 27 Oct 52.

and bunkers for enemy soldiers. Just then the Communists reacted with an even heavier supporting arms assault, which forced these few Marines to pull back with their platoon to the reverse slope of the ridge. To the right, about 250 yards away, the main body of Company I Marines occupied the reverse side of the hill, riding out the onslaught of artillery and mortar rounds while they waited for a lull before making the final dash to recapture the lost area of JAMESTOWN.

While Captain Harlan's company was exposed to this extremely heavy enemy artillery fire, another unit, B/1/7, was on the move from Lieutenant Colonel Dulacki's command post to the ridge to strike what was intended as a lethal blow to the Communist invaders. At 1932, Company B began its march forward. By midnight, the 1st Platoon was nearing its assault position close to the left flank of Company I of 3/1. Simultaneously, the 3d Platoon closed in on its jump-off point. The going was extremely difficult, complicated by a moonless night and the many shell craters that pockmarked the terrain. But at 0019, 28 October, the platoons mounted their assault, firing their rifles and machine guns, and hurling grenades to silence enemy automatic weapons and to reach dug-in Communist soldiers occupying the trenchline.

The Marine charge was met by a burst of small arms fire and a shower of grenades. Weapons supporting the Chinese defense were still very active. After a standoff of 90 minutes the Marines pulled back, calling on their mortars and artillery to lay precise fire concentrations on the trouble spots. The weapons also fired on enemy approach routes through Ronson and Warsaw. After this preparation, Company B again made an assault against the enemy, at 0340. This advance was contested vigorously by the Chinese, but their resistance this time was not lasting. Quickly B/1/7 Marines deployed throughout the entire area, and by 0600 the Hook was again in Marine hands.

Before the victors could permit themselves the luxury of a breathing spell, there were a number of critical tasks that demanded immediate attention. Defense of the MLR had to be quickly and securely shored up for a possible enemy counterattack. The newly rewon area had to be searched for Marines, both casualties and holdouts, and for Chinese diehards or wounded. The company had to be reorganized. In addition to these missions, there were two others, regaining Ronson and Warsaw. As it turned out, the duties were discharged nearly at the same time. COPs Ronson and Warsaw were reoccupied by the 7th Marines at 0630 and 0845, respectively, on 28 October.

In organizing the recaptured position, the Marines were hampered to some extent by a dense ground fog. Nevertheless, work still went ahead on these necessary tasks. Most of the Hook area was held by Company B; the western part of the 1/7 line, south of the Hook, was still manned by the platoon from Company A and one from Company C. The 1st Platoon of Company B quickly searched the retaken area of the MLR (except the caved in parts of the trenchline and bunkers, which were investigated later), but found no enemy soldiers. During the day, as Company B expanded its responsibility along the Marine main line, the platoons from A/1/7 and C/1/7 were relieved to rejoin their companies.[34] Supplies began to move in, once the permanency of the defense had been established.

Overview[35]

In evaluating the battle for the Hook, it would appear that the Chinese assault against Reno was merely a demonstration or feint. By making a sizable effort near the primary objective after the attack there was well under way, the Communists expected not to obscure the real target but rather to cause the Marines to hesitate in moving higher echelon reserves to influence the action at the Hook. It was to the credit of the ambush force that the Chinese ruse was unsuccessful.

Including losses from the Reno ambush, Marines estimated that the Chinese actions against that outpost cost the enemy 38 killed and 51 wounded. The COP defenders and the platoon that had surprised the enemy counted 22 dead Communist soldiers during and after the Reno action. Together with the Hook casualties,

---

[34] As a part of the reorganization, H/3/1 remained in the right sector, and Company C, of the Hook battalion, filled in the middle. Company A was in position on the friendly side of that part of the ridge held by Captain Byrum's Company C. During the afternoon of the 28th, I/3/1 and H/3/7 also left Lieutenant Colonel Dulacki's area to rejoin their parent organizations.

[35] Unless otherwise noted, the material in this section is derived from: 1stMarDiv ComdD, Oct 52; 1stMarDiv PIRs 734–735, 741, dtd 29–30 Oct 52, 5 Nov. 52.

Made in the USA
Monee, IL
04 November 2026

67377563R00125